ENTERTAINING FUTILITY

THE SEVENTH GENERATION
Survival, Sustainability, Sustenance in a New Nature

A WARDLAW BOOK

ENTERTAINING FUTILITY

DESPAIR AND HOPE IN THE TIME OF CLIMATE CHANGE

Andrew McMurry

TEXAS A&M UNIVERSITY PRESS • COLLEGE STATION, TEXAS

Copyright © 2018
by Andrew McMurry
All rights reserved
First edition

This paper meets the requirements
of ANSI/NISO Z39.48–1992 (Permanence of Paper).
Binding materials have been chosen for durability.
Manufactured in the United States of America

Library of Congress Cataloging-in-Publication Data

Names: McMurry, Andrew, author.
Title: Entertaining futility: despair and hope in the time of climate change
  / Andrew McMurry.
Description: First edition. | College Station: Texas A&M University Press,
  [2018] | Series: Seventh generation: survival, sustainability, sustenance
  in a new nature | Series: A Wardlaw book | Includes bibliographical
  references and index. |
Identifiers: LCCN 2018017930 (print) | LCCN 2018021448 (ebook) | ISBN
  9781623496869 (ebook) | ISBN 9781623496852 | ISBN 9781623496852
  (pbk. : alk. paper)
Subjects: LCSH: Environmentalism—Philosophy. | Climatic changes—Effect
  of human beings on. | Ecocriticism. | LCGFT: Essays.
Classification: LCC GE195 (ebook) | LCC GE195 .M455 2018 (print) | DDC
  304.2/8—dc23
LC record available at https://lccn.loc.gov/2018017930

Dedication

In memory of John Hitchcock McMurry (1921–2010)
& Mary Irene McMurry (1919–2016)

Contents

Foreword, by M. Jimmie Killingsworth	ix
Acknowledgments	xiii
The Slow Apocalypse Revisited, or The Supremes	1
The Green and the Brown	21
What Not to Do as the World Burns	37
The Denialists, or What If This Present Were the World's Last Night?	59
The Hopists, or A Sense of an Ending	75
The Decivilizationists, or Impure in a World Unpurged	95
iPod and World System	115
Entertaining Futility	137
The Narrow Corduroy Road to the Interior	165
Bibliography	191
Index	199

Foreword

This book is a collection of original essays with a common premise, a what-if premise such as you see in classic science fiction. What if a comet hit the earth? What if there were no gender? Andrew McMurry's premise asks, What if the worst is true, that the environmental crisis is not off in the future, but right here, right now? What if it were already too late? Drawing on wide-ranging scholarly and personal experience—from the study of literature, rhetoric, film, philosophy, history, and ecology, from camping and hiking and moving between city, suburbs, and woods—and writing with a rare and engaging eloquence, McMurry insists on raising questions that most folks would prefer to avoid. This is not a book that makes a point in the standard ways or proposes to solve problems. Such goals have been co-opted by the party of hope that has, in his view, led us astray again and again. In the old days, we might have called it prophesy. Now the best category might be what Richard Rorty calls "edifying philosophy." The main audience is the liberal academician or government agent or environmentalist who might take a pause from righteous indignation to entertain futility. Any reader will be edified by McMurry's broad reading and references if nothing else, but the hope is that the reader will see the world through new eyes and will be, if not changed as a person on a planet, then newly aware of the seriousness of the situation.

The common wisdom of recent years has suggested that the old environmentalism is too negative, too depressing; it undermines motivation and cramps progress. But here comes McMurry saying no. Perhaps the old environmentalism was too soft, too future-oriented, too progressive to grasp the depth of today's futility. The future that the environmentalists feared is on us. And yet we've lost the capacity to fear; we've talked ourselves out of it. The problem is more than environmental and societal; it's evolutionary in scope. To heck with sustainability, it's sustenance and survival we're talking about. That's why we need to entertain futility. We need to open the door and invite the unwanted guest into our air-conditioned and electronically enhanced parlors.

McMurry is an antihumanist scholar in the humanities. That is to say, he turns away from the trend of two-thousand-plus years of study that mire humanists in their focus on human products alone, particularly the study of words compounded by words, signs with the lightest possible reference to the earthly conditions of life, the dirt, the sea, the sun, the animal life that humankind is on the verge of destroying. To adapt the training of the humanist—the eloquence, the literary consciousness, the deep study, the sensitivity to language, the understanding of ethical reasoning and traditions of morality, the tracking of historical contexts and trends—to the work of ecology, and perhaps more immediately to bring the humanities up to speed on the findings of recent science, is, in many ways, an effort to think the unthinkable. Humanist scholarship is in decline not because it's dangerous, as many of us flatter ourselves into believing, but because it is irrelevant—twice or three times removed from the world, not inclined to attend to what's going on with the human relationship to nature, but instead interested only in nature and humanity as a theme of literature or a problem of thought, or worse yet, as a sentence in a theory of a theory of such themes and problems. The descent into irrelevance results not only from the insistence of the current university system on knowledge as a commercial product but also on the humanities themselves neglecting "the age-old critical and speculative functions of the university," as McMurry says. To engage these functions is to think the unthinkable, which must first begin "to imagine modes of life on Earth not rooted in utility and quarterly profit reports." To think such thoughts, the humanities must resist the complaints from outside the field that literature, philosophy, language, history, and rhetoric are useless or at least unprofitable and the insider problem of endless textualization that produces irrelevance (not necessarily the same thing as unprofitability). It's time to wake up before the humanities become a casualty that foretells the more general casualty of humanity. It's probably already too late.

What we call the unthinkable is in fact all too thinkable: we've likely destroyed the basis of life on earth. Rendering the thought as unthinkable allows us to continue in our destructive ways. Our films and stories, our products and our desires, our culture, are tuned to distract us from the thinkable. Go ahead, says McMurry, think it.

Read it and weep. Or read it and, in spite of yourself, enjoy a hysterical outburst of laughter, dark laughter, because *Entertaining Futility* summons a myriad of responses, most of which are resisted in general life. It's just possible that awakening these responses is a crucial step in getting past the distractions that allow the destruction to plow ahead. Again, the book is not a problem-solving book, but edifying in its purpose. If it adds even a single strand to our moral fiber, it is a very unusual work that deserves our close attention.

—M. Jimmie Killingsworth
 Series Editor

Acknowledgments

Jimmie Killingsworth prompted me to write this book and gave me sage advice along the way. Thanks, buddy. Big thanks to Ellen, Jack, and Drew for fun and laughter. Thanks to the Hooker Chemical Company circa 1977. Your wickedness set the long, low boil. Thanks to *Sitta canadensis*: paragon of restless energy, scattered purpose, and inverted viewpoint. Special thanks to Lake Michigan. Maybe not the greatest of the Great Lakes but definitely on the short list.

ENTERTAINING FUTILITY

Photo by author

The Slow Apocalypse Revisited, or The Supremes

> The startling calamity.
> What is the startling calamity?
> How will you comprehend
> what the startling calamity is?
> —Al-Qur'ān

Of the several things that must irk the universe about the human species, one is that we never like to acknowledge our transience. On the cosmic timeline humankind constitutes considerably less than a blip. If a dog year is seven human years, a human year is 180,000,000 universe years. Charles Lyell introduced us to deep geologic time in 1830, but the concept didn't take, and to this day humans fail to grasp the inconsequence of their collective existence, popularizing efforts by Carl Sagan and Neil deGrasse Tyson notwithstanding. *Billions and billions* (of years or stars or parsecs) aren't numbers a human brain can juggle happily. We can barely juggle now and then, as now converts to then in a jiffy. Memory holds on loosely to its contents. A great paradox is that every one of us thinks the current shape of things has been around time out of mind even as "twenty years ago" designates a forgotten backwater that lies under a cold fog. We're creatures of today and tomorrow, and History means whatever happened before the last Super Bowl. Faced with the concept of impermanence, humankind closes its eyes, puts its hands over its ears, and hums show tunes from *Annie*.

Twenty years ago, in a premillennial funk, I speculated that we were experiencing a sort of slow apocalypse, whereby our slide toward extinction was being revealed but at a clip that didn't puncture our comfortable presentism:

> Were you expecting the sun to wink out, the heavens to open, the beast loose upon the earth? Or maybe you imagined a Ragnarok of more cosmopolitan origins: nuclear war, bioengineered plagues, alien invasion, supernova. In any case, it's clear the last days are upon us, but given the laggardly pace at which this doomtime is proceeding we simply haven't yet grasped its contours. We adapt well to changes not sudden, swift and terrible, and just as we come to terms with the incremental decay of our own bodies and faculties, we learn to overlook the terminal events of our time as they unfold, gather, and concatenate in all their leisurely deadliness. We have wrongly expected the end of the world would provide the high drama we believe commensurate with our raging passions, our bold aspirations, and our central importance to the universe—we are worthy of a bang, not merely a whimper. And let me be blunt: By holding out for that noisy demise, we can pretend we haven't been expiring by inches for decades. (McMurry para. 1)

Then, obeying a questionable logic, I argued that this obliviousness to deep time's reduction of humanity's scope had something to do with the fact that human beings are short-timers on this planet:

> Clearly, this accommodation to the ongoing apocalypse is in some measure the result of our limited temporal perspective. In terms of recorded human history, the span of a few progressive centuries since our medieval torpor is

brief indeed. On the geological clock, the whole of *Homo sapiens*' rise and spread over the earth is but a few ticks of the second hand. Yet how seldom is this belatedness to the cosmic scene granted any significance! How, in our ephemerality, is it even comprehensible? Does a mayfly grasp that its lifetime lasts a day? From our blinkered, forward-looking orientation what happened a decade ago is already hazy, and the last century has become a bygone era. The only question asked of the past is, "What have you done for me lately?" Evidently, history is a nightmare from which we have awakened. Our neurological incapacity to hold in our minds with firmness and freshness anything but the near past and the now allows us to file away history as rapidly as we make it. Thus, absent a hail of ICBM's or seven angels with trumpets, the apocalypse can have been upon us for some time, may abide for another lifetime or more, and not until those final, tortured moments may it dawn on us at last that the wolf has long been at the door. (McMurry para. 2)

Today, much older but no wiser, I have not much changed my thinking on this matter—though the wolf has pushed his muzzle into the foyer and is drooling on the parquet floor. Climate change, which was on my naughty list then but not yet the main driver of all my work, is ratcheting up the velocity of the slow apocalypse. But what maybe needs to be spelled out, something I didn't bother with then, is the nature of this remarkably laid-back 'tude. Whence this stubborn immovability of humankind in the face of a mountain of evidence that we're here for a good time, not a long time? Back then I thought it was enough to observe, in my supercilious, scholarly way, that we were ignorant, self-satisfied beings about to receive a nasty comeuppance.

Our mollusk-like involution didn't need to be explained, just described.

Now I'm thinking a fuller accounting is called for. A slow apocalypse like ours spaces out moments of impressive cultural achievement with moments of sheer terror (for example, Harlem Renaissance and Operation Barbarossa). In the interstices, there is time for sober reflection. Minerva's owl may not actually fly at dusk but in the flat light of late afternoon. We're on the cusp of something, certainly a cascading worldwide environmental catastrophe, but maybe, too, a kind of *second enlightenment* in which we'll finally see, to our great consternation, that the long march of civilization has always been contained within a longer frogmarch toward oblivion.

Let's face it: The human species is a homeless creature pushing a grocery cart of dumb ideas piled so high they obscure the road ahead. On that cart rest, I suppose, some necessary illusions. But mostly what's been collected are junk-thoughts that we've grown fond of and can't bear to part with. Over the years I've cherry-picked from this psycho-rhetorical heap a few items that help account for our inurement to the slow apocalypse, this beast that hides in plain sight. What should we call them? I deplore the concept of *meme*, which tosses out whole histories and theories of ideology, discourse, cultural evolution, and cognition in favor of a dubious analogy between genes and catchy ideas. But I have less distaste for a concept that, though it may seem like a hypertrophied version of the meme, nevertheless holds a certain appeal when you're trying to bundle a lot of loose tendencies that seem in some obscure way to rest on the combined bases of biology, psychology, social norms, politics, and discursive formations. Rebecca Costa, in *The Watchman's Rattle: A Radical New Theory of Collapse*, defines a *supermeme* as "any belief, thought, or behavior that becomes so pervasive, so stubbornly embedded, that it contaminates or suppresses all other beliefs and behaviors in a society" (47). I like this definition, so I will borrow her idea, modifying the nomenclature slightly.

Here then, in reverse order of gravity, are the *Supremes* of human belief, thought, and behavior that mystify our relations with the ongoing planetary disaster and allow us to temporize in the face of catastrophe. The Supremes, to put it simply, are those dangerously powerful, mostly delusional, regrettably intractable, thought-ways that license us to believe we are apocalypse-proof.

8. THE CASSANDRA COMPLEX

In the Greek accounts, as everyone knows, Cassandra was a prophetess who could predict the future accurately but was disbelieved by all. Her name is frequently invoked to refer disparagingly to, for example, outspoken environmentalists or scientists who are actually eminently credible but nevertheless go largely unheeded by the vast majority of us. Al Gore and James Hansen or the team of scientists who produce the Intergovernmental Panel on Climate Change (IPCC) reports are cases in point. They've done yeoman service in communicating factual matters of existential import. Why then are they controversial? The original Cassandra was cursed by Apollo after she rebuffed his sexual advances. Her prophecies went for naught because the jilted god set it up that way. He made sure her fellow Trojans reviled her and thought she was nuts. No divine curse hangs over Gore or the IPCC; in fact, they have won Nobel Prizes. In principle, their pronouncements should have long ago galvanized peoples and nations into determined action. That they have not points to a more mundane interpretation of the Cassandra complex: Nobody likes know-it-alls, especially know-it-alls who are shouting the fun is over and the house is on fire. Such a person may well be onto something, but what's more pertinent is that she's a damned wet blanket. Among primates, signals about advancing dangers spread through the troop like wildfire. If the warning turns out to be a false alarm, so be it. The precautionary principle seems to hold sway, and feces aren't thrown at the offending chimp. No such luck for Al Gore.

7. THE OEDIPUS DEFENSE

As with the Cassandra story, everybody fixates on the sensational features of the Greek accounts of Oedipus—kills father, sleeps with mother—while ignoring what may be the more salient aspect of the tale: Oedipus rejects the truth of his family dynamic despite ample evidence in plain view all around him. His denialism epitomizes a profounder truth: Nobody wants to believe the worst about himself or his situation. We reshape our narratives to conform with our rosier self-portraits. In *Living in Denial: Climate Change, Emotions, and Everyday Life*, sociologist Kari Norgaard studies the reactions of the denizens of a Norwegian town to a particularly mild winter that is explicitly connected in the media to global warming. She finds that despite individuals' awareness of the clear linkages between their local experience and anthropogenic climate change (driven, in good measure, by petroleum production on which Norway's robust economy hinges), the town overwhelmingly engages in protective cognition, which is to say it *knows* the truth but it behaves as if it *doesn't*. The lessons learned in the Norwegian town, she claims, are directly transferable to the United States. She writes, "Through a framework of socially organized denial, our view shifts from one in which *understanding* of climate change and *caring* about ecological conditions and our human neighbors are in short supply to one whereby those qualities are acutely present but actively muted in order to protect individual identity and sense of empowerment and to maintain culturally produced conceptions of reality" (207). At the end of the day, sublimating the truth of our situation allows our quotidian self-conceptions and behaviors to proceed as before. Like Oedipus, we understand that things are rotten in Thebes, but we've got a kingdom to rule and can't spend the day blubbering.

6. THE LUCIFER GAMBIT

Satan is another mythic figure whose psychopathology speaks to our current immobility. In *Paradise Lost*, Milton created a not entirely unsympathetic rebel angel who rejects any authority other than his own. Satan, casting about self-servingly for justification for his war against Heaven, reflects to his fellow fallen that "we know no time when we were not as now; / Know none before us, self-begot, self-raised" (V 859–60). In other words, if somebody created us, if somebody is owed our allegiance, well, we don't recall that chap's name. As far as we know, we've always been here and have only ourselves to answer to, and so we must be guided by our own lights. "Our puissance is our own; our own right hand / Shall teach us highest deeds" (864–65). Lucifer's overconfidence could be due to a couple of strategies that lead the rest of us in a like way toward prideful pretensions of permanence. The first is an overweening faith in probabilities: If something has been happening for a long time—for example, you get out of bed each morning—it's tempting to extend this pattern into the future and assume events will continue as before. High probability is wonderfully comforting—right up to the night your heart explodes and you die in your sleep. The second strategy works on the basis of what rhetoricians call the fallacy of repeated assertion. Self-affirmation rituals are a good example. Those of us lacking in confidence are advised by our therapists to tell ourselves a benign lie—such as "I am a winner!" or "People love me!"—

over and over again until a protective crust develops over the fragile ego. We begin to believe our own press. An entire industry of personal growth and improvement has developed around this concept of repeated assertion. Once again, it's all very well and good to revel in your puissance and brandish your right hand—until the economy goes south and you're laid off, or your prickly personality finally drives away your last friend. Then we find, like Satan, that we were actually in God's hand the whole time, and He was dangling us very loosely over the infernal pit. All in all, what we should know—though we rarely learn it—is that strong self-confidence and good past performance are not sure predictors of the future. Better to view the past as a sampler of the kinds of bad things that will catch up to you sooner or later, especially if you act like they won't.

5. THE SUNNY SPARKLE SYNDROME

In the 1940s, the comic book hero Captain Marvel was a rival to Superman. Indeed, for many years his adventures outsold the latter's. But due to copyright infringement, the publisher Fawcett shelved "the Big Red Cheese" (as his archenemy Dr. Sivana dubbed him) for two decades. I first encountered him during the early 1970s, when he was purchased and rebooted by DC Comics. One of the appealing things about Captain Marvel was that his alter ego was a lad of my age, newsboy Billy Batson. Billy had only to speak the magic word, Shazam!, and a gaudy lightning bolt transformed him into the good Captain. He was endowed with the powers of great heroes of legend, their names encoded in the mystical invocation: **S**olomon, **H**ercules, **A**tlas, **Z**eus, **A**chilles, and **M**ercury. Anyway, it so happened that in the Captain Marvel universe was another boy with the delightful moniker "Sunny Sparkle." Sunny was invariably described as "the world's nicest guy." He was a walleyed, freckled, beaver-toothed blond with a bowl cut, possibly modeled after Jimmy Osmond. Anyone who met him could not resist giving him things: Rich old women stuffed his pockets with their jewels, average guys handed him their car keys, even burglars turned over their loot. Everybody loved Sunny Sparkle; he was a kind of reverse *objet petit a*, repository for the desired partial object. Bear with me now: I propose that the so-called human race constructs itself in the image of Sunny. Our species interprets its experience on this planet in ways that cast it as the universe's darling; indeed, it believes that the universe finds it irresistible. Simply put, we exhibit a deeply misplaced anthropocentrism that cannot be expunged. Despite its continuing efforts, the cold, unfeeling world described by science or naturalistic fiction just can't get a grip on human vanity and shake some sense into us. All evidence to the contrary, we humans possess a rock-solid faith that existence is *about us and for us*. Now, notice I do not claim that Narcissus should name the syndrome I'm describing (though narcissism covers a lot of this pathology, no question about that). The guileless Sunny Sparkle is the more apt patron saint, despite his lack of vanity, for the following reason: After Sunny received the unasked-for gifts, he turned them over, naïf that he was, to charity. Sunny was no unmitigated greed-head. And I think in an odd way humankind also fancies itself to be "giving back" to the universe. We are the center of things, God's chosen, so, yeah, sure, we can condescend to praise His good taste. That, at least, is the plotline of all the major religions, whereby adherents accept the benefits the godhead showers on them in exchange for worship, prayer, and obedience. Religion, in this view, is merely the formalized expression of our magnanimous self-absorption. Part of the human charm, in other words, is that while we believe the universe smiles on us, we try not to be ingrates about it.

Unfortunately, all that cosmic toadying serves only to reinforce and justify the underlying psychosis.

4. THE TOM SWIFT AND HIS BIG PLANETARY FIXATRON MANIA

At the same time I was into *Shazam!* comics I was racing through the Tom Swift Jr. adventure series, all thirty-three of them. These first appeared in the 1920s and, then, like the Hardy Boys mysteries, were reimagined for the more sophisticated sensibilities of the 1950s and 1960s. Racist stereotypes ("'Genmens, dinnah am serbed!' An old negro thrust his white-fringed head through the library door. 'An' it sho' am good!'" [Appleton, *Tom Swift and His Giant Telescope*]) were replaced with merely hackneyed ones ("'Mornin', buckaroos!' The chunky figure of Chow Winkler came into view. Formerly a chuck-wagon cook in Texas, Chow was now head chef on Tom's expeditions. As usual, a ten-gallon hat was perched on his balding head and he was stomping along in high-heeled boots" [Appleton, *Tom Swift and His Electronic Hydrolung*]). Unchanged were the technophilic plotlines, wherein Tom and his chum Bud Barkley foil rival scientists, hardened criminals, Red spies, and natural disasters by way of Tom's amazing inventions. No challenge was too mighty for the ol' techno-fix. What was true for Tom Swift Jr. was true for the nation in these can-do decades: Big Science and its armature, technology, held unlimited potential to create the dazzling future that could already be glimpsed in their contemporaneous successes. I remember as a child sitting rapt before TV shows like *Here Come the Seventies* and *Space 1999*. The distant year 2000 stood for a kind of zenith point in human civilization by which time all the emerging trends would have come together: moon colonies, asteroid mining, undersea farms, robot servants, and personal jetpacks. I suppose this Gernsbackian notion has been tempered, chastened even, in the minds of most of us. Some of that future now looks to be out of reach, owing to the physical limits of matter, space, and the human organism. It takes a *lot* of expensive energy to escape Earth's gravity well or drill a hole to its molten core. Tom's "repeletron skyway" and "cosmotron express" remain exactly what they were in 1960: total flights of fancy. Yet at the level of practical consciousness, what could be more future-affirming than the continual rollout of new consumer gadgets and services? In this volcano of desire and fulfillment, we all become indoctrinated into an enduring albeit shallow faith: *Something* will come along to calm the whirlwinds that humans have released on the earth. It's very difficult to combat this sort of magical thinking, even when it gets taken down a peg or two, what with your Bhopals and Fukashimas, your *Challenger*s and *Exxon Valdez*es. In *Earthmasters: The Dawn of the Age of Climate Engineering*, Clive Hamilton writes, "Prometheans rule. Over three centuries of advance, displaced workers, romantic poets, dismayed clerics and far-seeing ecologists put up resistance; all sooner or later were crushed. Who can hold back such a force?" (209). Who, indeed, and, as far as almost everybody is concerned, why hold it back at all? As climate change begins to degrade food and water resources even in the developed world, more and more of us will find ourselves fluent in the finer points of *solar radiation management*, *scrubbing towers*, and *enhanced weathering*. We'll have placed our hopes in the Next Big Fix because, well, banking on what's next is just the way we roll.

3. THE TOM SAWYER WHITEWASH

For the nineteenth century, the parable of Aunt Polly's fence nicely illustrated Young America's entrepreneurial spirit. In conniving to get his friends to do his painting chore for him, Tom "had discovered a great law of human action, without knowing it—namely, that in order to make a man or a boy covet a thing, it is only necessary to make the thing difficult to attain" (32). Today, the creation and curation of desire are more crucial to the economy than timber and potash, and inducing someone to do something for you while simultaneously charging them a toll is a wet dream of capitalism. What we don't often realize is that this dream comes true every day. When a corporation strip-mines a landscape for coal, that's Tom Sawyering of the first water. How so? The workers, communities, and natural environment—this assemblage from which the mining firm draws its support and resources—does the dirty work and pays with its socioecological health. Petroleum, manufacturing, agriculture: All the major players on the global scene are in the business of creating demand and off-loading the costs. They draw their basic stuffs from the planetary commons, work it up here or there depending on who will bid to labor on it most cheaply, then move it from one part of the world to another looking for the most willing buyers, at every step spewing unaccounted volumes of carbon into the atmosphere, leaving waste and want at the end of it all. A codfish scooped up by a factory trawler in international waters off Newfoundland will be shipped to China for processing, then back to Saint John's for insertion into a McDonald's sandwich, and eaten by an unemployed inshore fisherman. A 20,000-mile round trip for a half-pound fillet. In effect, our transformation of the planet for the profit of a few displaces the costs onto all of us. We think we are more or less fairly compensated for our labor and nature's sacrifices, but we are paid a pittance and we pay out a fortune. There was a time when all this displacement didn't go down well. Movements arose to recover the value stolen from the labor process. Then the pendulum swung again. Nowadays economic well-being is pegged to the system as a whole, only incidentally to the individuals who serve it. Expansion is the key indicator of system health. That this expansion comes at the cost of the planetary life matrix is tricky for individuals to bring into focus, especially when all their rewarded efforts are in the areas of production and consumption. Heaven is the perpetual swelling of the economy's operations, and, as Peter Sloterdijk notes, "Hell is the impossibility of expanding" (237). Here we have the second aspect of the whitewash: the fully operationalized dogma that perpetual growth on a finite planet is not only possible but also inevitable. This Supreme can be distilled to the old truism, "There's no such thing as a free lunch," to which must now be added the qualifier "but they sure do taste good!"

2. THE POLLYANNA IDEOLOGEME

Probably all these Supremes I've mentioned so far could be modified by the same descriptor: optimistic. Among the concepts Modernity disgorged was a view that the world was getting better, that the people and stuff in it were on the upslope. Conditions for improved existence were being laid out and roughed in, and, along with the material benefits brought by science, medicine, and technology, something even rarer and higher was abuilding. That something was human moral agency: We were developing just laws and governance, sloughing off older bigotries and biases, and, generally, refining ourselves into beings of purer quality. History was working to produce, if not yet a utopia, then a place that would be, in the fullness of time, a very desirable address. If optimism drove this project or was merely a by-product, I don't know. All that concerns me is that optimism was waxing, replacing the more fatalistic attitudes that had animated the inner lives of our predecessors. We propelled ourselves by hook or by crook into a kind of Pollyannaish reality in which even the darkest cloud had a silver lining. ("That world war was surely awful, but it did give us tanks and flamethrowers.") This new construction of fate led folks to be bright-eyed about their personal trajectories rather than lachrymose. The Pollyanna principle, according to its identifiers, "states that pleasantness predominates; pleasant items are processed more accurately and efficiently than unpleasant or neutral items" (Matlin and Stang 3). When it comes to future states of happiness, folks tend to believe their own lives are charmed—over and against immiserating trends on the national scene: "Even if we fear the worst, it seems we may expect the best" (163). Pollyanna, by the way, was a fictional character created by Eleanor Porter in her 1913 novel of the same name. Famously played by Hayley Mills in the 1960 Disney movie, Pollyanna was transformed into a synonym for a foolish, unwavering optimist. She came by her optimism honestly, a function of playing what her father had dubbed the *glad game* whenever the world dealt her a losing hand. The glad game is exemplified by a story she tells her nurse about a potentially distressing gift that arrived in a "missionary barrel" for her and her father, Reverend Whittier: "You see I'd wanted a doll, and father had written them so; but when the barrel came the lady wrote that there hadn't any dolls come in, but the little crutches had." The nurse, mystified, says, "Well, goodness me! I can't see anythin' ter be glad about—gettin' a pair of crutches when you wanted a doll!" Pollyanna, delighted, replies, "Goosey! Why, just be glad because you don't—NEED—'EM!" We have here a cornpone version of Leibniz's theodicy, in which the great optimist famously explained that a wise and generous God has ensured that all is for the best in this best of all possible worlds. The great Lisbon earthquake of 1755 that obliterated 40,000 lives was scarcely comprehensible in its enormity, shaking Europeans' faith in the divine plan. But given time and distance, argued Leibniz, we would come to know its karmic benefit. In other words, God's got it covered. Our contemporary Pollyannaism isn't so lurid. After all, we are not dopes. We ground our optimism in good and sound reasons, all security-checked and verified—something to do with the unlimited supply of human ingenuity, the invisible hand, and the principle of resource substitution. And if that's not enough, if you still feel the need for emotional and spiritual bucking-up, you don't have to look far. Mary Baker Eddy showed there was power in prayer and Norman Vincent Peale in positive thinking. Tony Robbins added the concept of peak performance and Oprah self-esteem. Bill Clinton touted hope and Barack Obama audacity. Homer J. Simpson wondered if there was anything donuts couldn't do and Donald J. Trump if there was anything worthwhile he couldn't undo. Here's the gist: According to the North American cultural consensus, nobody ever solves a problem by dwelling on the downside. Keep thinking the good thoughts and the world will spin on your personal axis.

1. THE CTHULHU MYTHOS

A culture can't embrace optimism without rejecting pessimism. Unlike the other Supremes, this one is observed mostly in the form of its negation. It can be understood as a no-fly zone for thinking and speaking or as a mental setting that in our culture is all too frequently flipped to the off position. In my view, the problem with pessimism today is not that there's too much of it but rather that there's not nearly enough. Nobody seems to like catastrophizers, not even other catastrophizers. In fact, pessimism shouldn't be understood as a matter of gloominess or cynicism but of philosophical and intellectual honesty. For my purposes, pessimism is akin to a sense of terrible wonder, "an affirmation of life in the teeth of its limits" (Lasch 308). It is, paraphrasing Schopenhauer, the recognition that the universe does not exist to make us happy—a kind of realpolitik applied to the soul. The optimist says the glass is half full, the pessimist that it's half empty. (The engineer, by the way, says the glass was built too large, which for my money effectively aligns him with the optimist in that he's saying pessimism can be designed out of existence.) The Italian political theorist Antonio Gramsci advised us to espouse a pessimism of the intellect and an optimism of the will. But in our culture, we like our optimism neat: Both will *and* intellect are left to marinate in optimism's bland waters. All in all, the failure of pessimism to make any great inroads in the Pollyannaish culture means that the unthinkable does not get thought—not because it can't be thought but because it is uncivil to do so. We cannot bring ourselves to scrutinize the unnameable, the void, the monstrous empire of dubiety that is heaving into view. We do not agree with Oswald Spengler, an unreserved miserabilist, who noted "Only dreamers believe that there is a way out. Optimism is *cowardice*" (104). We fail to acknowledge that pessimism might be the stance of the brave and the clear-eyed; rather, we relegate it to a malaise of the mind, some sort temperamental bile that one is advised to expel ASAP. Yet I like to think a bracing pessimism is the essential corrective to the straitjacket of optimism, with its mandatory faith in progress and human exceptionalism. A pessimist looks at our worldwide industrial civilization and sees a giant house of cards, all the various rooms and extensions bound in one precarious arrangement. If a single section fails, the whole edifice comes tumbling down. But the house of cards is an unattractive analogy in this cosmopolitan age: it's archaic, nondigital, and homely. Indeed, the optimist rejects this analogy out of hand. Instead, he compares civilization to the internet: a vast, complex system with built-in firewalls and work-arounds. If one section fails, the organization contains enough redundancies to repair itself and continue on more or less as before. A very attractive portrait of resilience, to be sure, one that jibes with the latest and coolest mental schemas, informed as they are by computer architecture, the social network, and postmodern relativism. But which analogy is correct? Both and neither. The point is that the optimist works on the Panglossian principle, whether he knows it or not, that all is for the best in this best of all possible worlds. He is confident civilization will be rerouted around the little glitches that periodically pop up. The pessimist prefers to proceed on the basis that the worst possible scenario is the likeliest, and that when the shock comes, we should expect no help. She adheres not just to the precautionary principle but to what might be called the collapse principle: that the assured outcome for all systems—cards and webs alike—is dissolution. All she can do—all any of us can do—is to try to delay the collapse by combating the immoderate confidence that makes it ever more likely. The pessimist lives in a mine shaft and spends her time lining it with timbers. Not surprisingly, few of us want to visit. But the time has come to *cowboy up*. The age of optimism is over. The many-tentacled horror of life-in-the-Anthropocene is on us, and "loathsomeness waits and dreams in the deep, and decay spreads over the tottering cities of men" (Lovecraft 157).

Messenger-killing, head-in-sandism, fallacious endurability, cosmic apple-polishing, techno-deference, expansionist ideation, glad-gaming, and the gloom prohibition: these are my eight deadly sins of the Anthropocene. To this octet of Supremes we could add in any number of *social proofs*: herd instinct, groupthink, Young Werther effect, among others. When confirmation bias, base rate neglect, hedonic psychology, loss aversion, apophenia, shifting baseline syndrome, and about nineteen other sorts of popular errors and delusions are included, one can only throw up one's hands and mutter, in the idiom of our time that so well expresses the impotence of nuance before vulgar reality, *it is what it is*. In fact, the list of human foibles and failings is so extensive it's not much of an exaggeration to say that when it comes to our remarkable penchant for tuning out the slow apocalypse there are only a few human features—fondness for pets, sunsets, and naps come to mind—that don't contribute to our wide-ranging doltishness.

I will close with the same questions I posed in my earlier 1996 essay. My elucidation of the Supremes does not answer these questions, not by a long shot. But now I've provided a better map of the psycho-rhetorical briar patch we're struggling to exit. It's become too easy to blame all our problems on structures and institutions; doing so takes the heat off us, the beings who built them. But you have to wonder: What the hell is wrong with us that we can't begin to repair what we've spent so long screwing up? You'd think at least we'd make a stronger effort. I've heard it said that humans are not insane creatures; we just do things that are insane. Perhaps whatever sort of sanity we imagine we possess just doesn't work anymore, not on a planet that we've spent millennia sanitizing. Perhaps Theodor Adorno and Max Horkheimer were right when they observed, "the wholly enlightened earth is radiant with triumphant calamity" (1).

So I ask again: Who says the human presence on this earth was ever sustainable? Why do we continue to believe so strongly in our competency to manage the risks we compound daily? Where is this secret heart of history we trust has been beating? What precisely leads us to believe our world is not perishing? Why isn't this the Apocalypse?

Photo by author

The Green and the Brown

> Only one thing can save us. We have to increase our mastery of the world. All this damage has come about through our conquest of the world, but we have to go on conquering it until our rule is absolute. Then, when we're in complete control, everything will be fine.
>
> —Daniel Quinn, *Ishmael*

Now there's a sentiment most of us can wrap our minds around: that we humans have so botched our tenancy on this planet that there's nothing for it but to stick with the mad plan that got us this far. Halfway lost is halfway found. What will be there to greet us when we stumble bloodied and dazed out of the woods—well, we're not quite sure. But worrying about the future doesn't get us anywhere, never has, never will. Quarterly profit statements are a good indicator of our fixation on the here and now: Will a business survive even into the next month if it's worried about the next hundred years? Obviously not, and so our exertions must be directed toward taking care of today. Tomorrow will take care of itself. Far from espousing the Iroquoian notion that we should deliberate in light of the seventh generation to follow us, our gambit is that the seventh generation had damn well better be advanced enough to handle the problems we intend to dump at its feet.

Has it always been this way, a feckless primate manhandling the planet, its creatures, its future? Well, it seems that for as long as humans have been upright, we've been very good at taking. But when the species was young, it could not take very much: some wood for the fire, a handful of shells for the shaman's pouch, a few feathers for the chieftain's headdress. There were not many of us—perhaps a million at most, scattered about the habitable parts of the globe—and the world was a cornucopia. What we took was not missed. For tens of thousands of years, nothing much changed. But over the last few millennia, the scale of operations broadened; our numbers ballooned and our means of altering physical reality grew commensurately. Indeed, in just the last two or three hundred years, humankind has seemed to grow bigger than the world itself, our power multiplied by technology to the point that we can only be compared to the other grand forces of nature: the tides, the shifting plates, the climates. By the time we noticed the stillness on the plains and the silence in the forests it was much too late to withdraw our bloody hands. The bison had been skinned and the parakeets plucked, the ancient hills leveled and the salmon streams buried. The sky itself was thickened and the shallow seas scraped clean. We had reached the edge of the world, and the void dropped away before us just as the stories foretold. There was simply not enough world left to satisfy seven billion souls' ambitions, perhaps not even their barest needs.

The world is a mysterious place, and it comes with no instructions. "Life is an experimental journey undertaken involuntarily" (Pessoa 74). The human predicament has always been at the center of scholarly thought. The same existential questions have been pondered again and again, across all divisions of knowledge, by all cultures that left a record. The observation of the remotest star and the innermost recesses of the mind, the labor of poets and engineers, the tracking of amphibian populations and the dollar—all converge on the basic unanswerables of humane inquiry. Who are we? Where are we going? What shall we do? Yet it's possible, even reasonable, to argue that answers to these questions are pursued so vigorously only because we lack a durable, satisfying framework for living in the here and now; in its absence, we are compelled to imagine that if we could just amass a bigger collection of facts we might thereby soothe the ache of our mortal condition and bring into being that better world our investigations hint might be possible. In the meantime, our efforts at mastery continue to mutilate the earth, almost as if to secure the future we intended to destroy the present.

In our honest moments, we recognize this folly; we know the collapse is coming. The Malthusian calculus was never disproved, only delayed. The strains on the biosphere never ease, only tighten. The real problem, of course, is not with our knowhow; if all that was required were reason and technique, we could begin repair of the mess tomorrow. But the problem is thornier than that. Every one of us intuits, deep down, that we are not quite at peace with the world. We seem constitutionally dissatisfied with it. It's never been possible for a man to sit quietly on a rock and leave the world alone. Soon enough he'll find a weed to pull or a bug to squash. We're the ones who have to be about our father's business, no matter what it is and no matter where it leads.

Perhaps on the road to becoming human, something was removed from us. Or perhaps something's been added. Either way, we no longer fit the earth as other animals do; we feel compelled to make it fit us. We understand what we're up against—ourselves—but we fail to leverage our comprehension into anything but more grasping. Heedless of the terrible costs of unbridled desire, we "want and want and want," as Joy Williams writes, believing in ourselves "excessively" while no longer believing in Nature, our eyes glazing over as we "travel life's highway past all the crushed animals and the Big Gulp cups" (3). And whatever it is that we want, we're willing to pull the whole edifice of life down on our heads to get it. We are the new Easter Islanders, cutting down the last tree to roll the last stone head up the last unmonumented slope, smug in the knowledge that though we've doomed ourselves, at least by god we're going to finish what we started.

According to Daniel Quinn's Ishmael (who happens, by the way, to be a 1,000-pound telepathic gorilla with an outsider's insight into the human condition), the struggle against our own imperious gluttony repeats the pattern of a long-standing cultural battle between what Quinn calls the Takers and their counterparts, the Leavers. Takers created all the great civilizations the world has known, including our own. They settle down, make a mark on the world, and they strive for permanence. Leavers, by contrast, deal with the exigencies of living in a gentler way. Such cultures are preagricultural; they neither settle in one place nor generate the surpluses of food that can tide them over while they pursue other ends. Leavers move from food source to food source, always keeping their populations low and only lightly taxing the ecosystem. Nomads, hunter-gatherers, subsistence herders, pastoralists: these so-called primitives enacted a different cultural story than that of the Takers, which claims that the world belongs to man. The Leavers story, by contrast, says that, "man belongs to the world" (240). It's a standard progressivist mistake, by the way, to assume that the nomadic Leavers welcomed their conversion to the sendentist Taker strategy. It seems more likely that Leavers recognized the civilization trap, for "there is massive evidence of determined resistance by mobile peoples everywhere to permanent settlement, even under relatively favorable circumstances" (Scott, *Against* 8).

Needless to say, the competition between Leaver cultures and Taker cultures was settled long ago, notably in the fertile regions of the Middle East, where the ancient Semitic herds peoples were displaced by the Taker agriculturalists, who would eventually give rise to the great civilizations in Mesopotamia and along the Indus and Nile. There are very few of these Leaver cultures left, and almost none are intact: a few Bushmen of the Kalahari, some uncontacted Amazonian tribes, the Spinifex of the Great Australian Desert, the island-bound North Sentinelese. We don't know a great deal about Leaver history because in a sense they left no history. Humans, living day to day, in harmony with one another and with the world, rehearse the same story over and over again. Linear history only begins when the Taker agricultural strategy interrupts or destroys the endless, repetitive cycle of days: the wanderers settle down, create granaries and storehouses, towns and cities, temples and roads, markets and money, owners and workers, weapons and wars. Then there are many stories to tell.

If we accept this frame tale for the origins of humanity's dysfunction—and why not? it's as good as any—we still have the problem of explaining why we have not managed to overcome our grasping Taker mentality in the last few hundred years, as we proceeded through a remarkable period of human flowering, known variously as the Enlightenment, Modernity, or Age of Reason. It was during this epoch that humankind—or rather, European humankind—came to understand itself and its surround-world in ways that opened up new possibilities for care and concern vis-à-vis one another and nature. Driven by scientific, technological, social, and political advances, Euro-man was able to assess his existence in all of its dimensions, for good or for ill, and the Taker mentality that ruled for so long was brought into sharp relief. It wasn't abandoned—far from it—but the disadvantages of treating the world as a resource basket and a waste receptacle began to emerge. As the continents and oceans were mapped, assayed, and commoditized, it dawned on some that the planet was finite, and that humans could have profoundly negative effects on it. Were some refinements to our approach called for? They were. Yet if we look around us now, we see that regardless of those refinements, the impact of humankind on the planet is undiminished. It falls heavier each year, each day, each minute. Simply put, it was the Takers who set down the fundamental principles of what I will call the brown vision of the world—marked by resource exploitation, intensive cultivation, unlimited population and economic growth regardless of cost to the planet—and no green alternative has managed to budge more than a handful of us off those principles. Think of the many historically noted but practically inconsequential crypto-Leavers: the Diggers, the Mennonites, the Shakers, the Fourierists, the hippies, the bioregionalists, the vegans, the animal people, the Wall Street Occupiers! All of these groups were and are regarded suspiciously, suckers off the main stem of history, doomed to the margins as curiosities, laughingstocks or, worse, perversions of a full-blooded humanity.

The lesson here is that a reduction of human suzerainty on the planet is not culturally viable; we define ourselves as the earth's lords and masters, and we guard our dominion jealously. The first commandment of humankind might as well be "thou shalt never willingly scale back." Rather, *Homo colossus* must go forth and multiply its power across the entire range of material existence. Do we place any limits on our ambitions? Only one, which is that we will limit any discussion of human limits. We must proceed through to full Modernity, with technological, managerial, and economic upgrades providing the means by which our admittedly heavy footprint will somehow, magically, be rendered tolerable. The phrase *sustainable development* describes this impossible dream, whereby our massive combined impacts are imagined to lessen even as the sum our activities increases without end. It's no wonder we're able to hold such an obvious biophysical contradiction in mind. According to Lionel Tiger, we're hardwired optimists: "There is evidence to think there was an evolutionary advantage gained by people who think well of the future or of their immediate prospects" (20). As the great British-Polish sociologist Zygmunt Bauman puts it, "All human cultures can be decoded as ingenious contraptions calculated to make life with the awareness of mortality livable" (31). It stands to reason, then, that when we place bets on the future (for example, by calculating that some techno-fix will come along to mitigate the effects of pumping 20 gigatons of CO_2 into the atmosphere each year) we are predisposed to assume the deck is stacked in our favor. So we'll take the long odds. No guts no glory. There is everything to gain and only house money to lose.

There's something deeply unsettling about an Enlightenment that turns a green and blue world into a brown one. It is the nightmare into which we've plunged but which most of us do not distinguish from a beautiful dream. Max Horkheimer, who with Theodor Adorno literally wrote the book on the perverse reasoning behind the Enlightenment project, puts his finger on at least one of the paradoxes when he notes that the "domination of nature involves the domination of man" (64). That tight connection between what we do to our surround-world and what we do to one another can be described as a negative feedback loop: The more we are willing to subdue nature the more we are willing to subdue one another the more we are willing to subdue nature. It goes without saying that this hideous cycle must be interrupted, broken, foiled. Yet everywhere it gains in force and scale.

Homo homini lupus est. As man continues to be the wolf to man, he deals out death to the nonhuman world. One might even say that the modern rationalization promotes death even as it claims to be enhancing human life. In addition to the tens of billions of terrestrial food animals killed each year and the trillions of marine organisms, we now contemplate a generalized Sixth Extinction, the conditions for which have been locked in by our voracious mentality. Gary Snyder focuses on this pathology in his poem "One Should Not Talk to a Skilled Hunter about What Is Forbidden by the Buddha":

A gray fox, nine pounds three ounces.
$39^{5}/_{8}$" long with tail.
Peeling skin back (Kai
reminded us to chant the Shingyo first)
cold pelt. crinkle; and musky smell
mixed with dead-body odor starting.

Stomach content: a whole ground squirrel well chewed
plus one lizard foot
and somewhere from inside the ground squirrel
a bit of aluminum foil.

The secret.
and the secret hidden deep in that. (66)

Beginning with the close, forensic examination of the dead fox—an anatomy lesson that harks back to the early moments of scientific medicine—and through the quick summary of the fox's food web, Snyder traces the standard lineaments of the modern method: observation, classification, quantification, inference. But then the poem turns to the anomalous bit of foil, the Reynolds Wrap that must have held its razor edge through the chewing and digestion, a tiny blade to slice open the lining of the fox's stomach. The internal bleeding, the slow, incomprehensible death that overtakes him from within: that's the secret. And the secret hidden inside the secret? That humankind built this blade in the smithy of our great Enlightenment. Knowledge given form. Like most of our modern products, it is a casual thing, to be cast aside, forgotten by us—but never forgotten by nature. One of the four principles of ecology as famously stated by Barry Commoner: "Everything must go somewhere . . . in nature there is no such thing as 'waste.' . . . Nothing goes away; it is simply transferred from place to place" (36–37). That bit of foil you used to cover a plate of leftovers winds up in somebody else's belly or bloodstream. So it goes with your computer, your car, your diapers, your dishwater, your Glade Solid, your Tom's of Maine. Your wifi, your mercury, your fuel rods. Your breath. Your body. Your coffin. All the effluvium in the spinning, sucking modern gyre vomits forth into nature, which isn't just around you but is in you—*is* you—and so, literally, as Horkheimer says, the "domination becomes 'internalized' for domination's sake" (64).

And then there is the story of the monkey and the gorilla. Presented by the researcher with a jar full of nuts, the monkey puts his hand in the jar, closes his fingers around the nuts, and then can't get his fist back out. He's stuck. He can't remove the nuts but he refuses to let go of them, growing more and more frantic. By contrast, the gorilla, when it's clear he can't get his fist past the bottleneck, calmly releases the nuts. He pauses for a moment, then flips the jar over and pours the nuts into his palm. The story is meant to tout the remarkable intelligence of the great apes, our nearest cousins. But that interpretation is just another self-flattering gesture toward human superiority. To me the anecdote illustrates another point: Humans have no problem fucking around with even the smartest animals. From worm to rat to elephant, living things aren't alive for themselves. They're here to edify, entertain, and equip us. "Only man's soul can be saved; animals have but the right to suffer" (Horkheimer 71).

Human chauvinism is embedded in our everyday discourse, and we are hard-pressed to form sentences that do not betray our self-importance. Here, for example, is an axiom from a popular introductory textbook on human ecology: "The ecosystem provides services to the social system by moving materials, energy and information to the social system to meet people's needs" (Marten 2). A student reading this claim might be forgiven for imagining that nature's sole ambition is to be our unpaid gofer. Although the book wishes to challenge the dominant, exploitative paradigm of development, it speaks with same forked tongue as the conceptual regime it opposes. Another scientific paper warns us about the defaunation of the oceans, noting, "The most conspicuous service that marine fauna make to society is the contribution of their own bodies to global diets" (McCauley et al. 250). Well, I suppose that supreme sacrifice will be better appreciated when the final catch is soon scooped from the dying seas. But lest we imagine only as protein do marine fauna express their obligations to us, keep in mind that "a diverse array of nonconsumptive services are also conferred to humanity from ocean animals, ranging from carbon storage that is facilitated by whales and sea otters to regional cloud formation that appears to be stimulated by coral emissions." Next time it rains on your beach vacation, blame the polyps. Such is the way of most of our discourse. It has evolved, not surprisingly, to complement our sense of cosmic centrality and entitlement. In effect, humanity is the action star in all the scripts that are churned out by *Hollywood sapiens*; the cameras don't even start rolling until we walk on to the set.

Well, as Petronius put it, the world wants to be deceived. Or, as Mark Twain observed, denial ain't just a river in Egypt. But as with most addicts, we suspect we've got a problem. Sometimes, we're quite sure of it, as when our leaders and their surrogates, gathering in Davos, Camp David, or at the latest G7, G8, or G20 junket, scare us a little bit with their talk of economic downturns, pandemics, and terror states. We begin to wonder if the doom-and-gloom crowd might not have a point: Maybe we should be directing the planetary show with a bit more savvy. So we summon up the stern words and bracing ideas, hoping that they might somehow remove the monkey from our back, the monkey being this terminal arrogance that causes us to equate unmitigated disaster with progress. We are particularly keen to hear from those who saw the train wreck coming and offered a means to avoid it. Some of us, for example, will look to Henry David Thoreau, who sought voluntary simplicity in the woods outside of Concord, Massachusetts. Thoreau, grieving from the horrible death of his brother from lockjaw, built a ten-by-twelve-foot cabin of salvaged lumber near the shore of a small lake, and there began a healing process that eventually became one of the great self-help books of all time, *Walden*. In that book, some believe, are all the tools we need to cure our minds and get right with the nature.

Elsewhere, Thoreau says that "in wildness is the preservation of the world" (*Natural* 112). Furthermore, he claims that "life consists with wildness" (114) and that he would have "a wildness whose glance no civilization could endure" (113). What does he mean by that? Even in his time, there wasn't a whole lot of wilderness left. But, wait: he said "wildness," not "wilderness." Two distinct, albeit related, items. It's not wild places he requires (though he was fond of those to be sure); it's more a state of mind he asks us to adopt, a freedom from constraints imposed by manners, civility, and the market. He believes that the "savage" lurks within us, and we need to let him out. We crave the solace of the swamp, the marrow of the woodchuck, the liquor of the pine needle. We will "import wildness into the cities at any price" (112). Thoreau challenges us to give in to our hidden craving, not the grasping, self-defeating, self-loathing hankering for culture's baubles and blandishments but the primal call of the jungle, that interior wilderness of total uncivility that beats in our Paleolithic breast. He would no doubt concur with Horkheimer about the bargain of Modernity, that in order to subjugate nonhuman nature, man first had to "subjugate nature in himself" (64).

And that was a bad choice. By putting nature in chains, we unknowingly put our best selves under lock and key as well. For those who would like us to follow the Thoreauvian path, then, the first step is simply to recognize our enslavement to the false god of civilization. Only then can we begin a process of *rewilding*, a selective restoration of our lost connection with the world around us. And once we peel away the artifice and prosthetics that color all our relationships with one another and the object world, well, anything is possible. We could, for example, end the application of industrial-age, assembly-line processes to the growing of food or the education of children; we could reconfigure our cities and towns as communities of walkers instead of communities of cars; we could live between dawn and dusk and return the stars to the sky; we could become a culture of makers instead of consumers. We could live according to the seasons instead of the financial cycles, and we could direct our energy, time, values, and spirit toward inhabiting the world rather than toward rendering it uninhabitable.

Well, you might counter that Thoreau and his ilk ask a lot of us. We've come too far on the upper slopes of Modernity to roll back our gains. Sure, a canoe trip or walk through the woods is a healthy pastime, but to devolve into a raging wolfman sounds more like the plot of a horror movie than a realistic solution to urban angst or global warming. We need the opposite of wildness, you might argue, a fuller tameness, a more thorough suppression of our animal instincts and those unstructured urges that give rise to unhappiness, unrest, crime, pollution, and war. You might call for a tightening of our reasoning mask, a rededication to the principles of prudence, good judgment, and sobriety. Forward, always forward. You might say that even if civilization is sort of an airplane, and even if the flight is bumpy and the pilot and copilot seem to be sleeping or stoned, and even if the rivets are popping out and the airframe is twisting, and even if the flight attendants seem to be spending most of their time up in business class serving drinks to the well-heeled, well, running up and down the aisles shouting passages from *Walden* isn't going to help, now is it? You might say, let's be serious people for a minute, enlightened people: We need to fix this thing, mid-flight if we have to, and get it back on course. Maybe it is on course anyway, and all this turbulence is just a patch of rough air. You might say, don't give up on hope, hope is what got us here, and if our hope is audacious enough then how can we fail? Or you might say, thinking now well and truly outside the box, OK, let's play it safe, let's bring this machine to a soft landing and send for the people who designed it, the experts, and get them to tune it up. Only when we're absolutely assured of its soundness will we press on, with patience and good cheer, toward our destination, the awaiting utopia. Or you might even be one of those who says, wait a minute, what are you complaining about anyway? I didn't feel any bumps, and my pretzels and tomato juice are right here on my tray, and by the way I need to take a nap, so would you please keep it down? Enjoy the flight.

You might say any or all of those things, and that would mean on the one hand that you've missed the whole point of my discussion but on the other that you understood it all too well, that you realized you and I are passengers on this craft, engrossed in free movies and strapped to the seats, chained to them, to tell the truth, and that hope, prayer, and faith in the unseen engineers, wherever they are, is about all we've got left.

I was in our university research library the other day, and I heard a young woman say to a young man, "Originally I wanted to do animal rights. But there's no money in animal rights." I knew what she meant. There's no money in saving animals, let alone money for researching how to save them. The real money is in killing animals, and finding new ways to eat them. There's also no money in saving trees. The money is in cutting them down and finding new ways to turn them into engineered flooring. And there's no money in saving you from your own desires. The money is in making you turn away from what you already have toward all the things you don't. And, finally, there's no money in the facts. The money is in making the facts support money and those who possess it in abundance.

But is there no money in saving the planet, and ourselves? It would be pretty to think so, though greening the economy is about as achievable as isolating phlogiston or capturing Sasquatch. I do know there's money in selling the planet short, which is actually a technical term. To stock traders, it means betting that over time the value of something will decline, allowing them to profit from the downturn. The short trade is a metaphor for the broader program of which we're all a part. If you have ever come across a book with a title like "50 Simple Things You Can Do to Save the Planet" or "365 Ways to Live Green" or something along those lines, just keep in mind that each day, not in 50 or 365 but in 10,000 different ways of which you are mostly unaware, you yourself are selling the planet short. You are borrowing against a tomorrow you intuit will have less value than today. So I would say that between choosing the green future and the other, there is only our collective will. And at this point, it's become quite clear that our collective will has already chosen the brown.

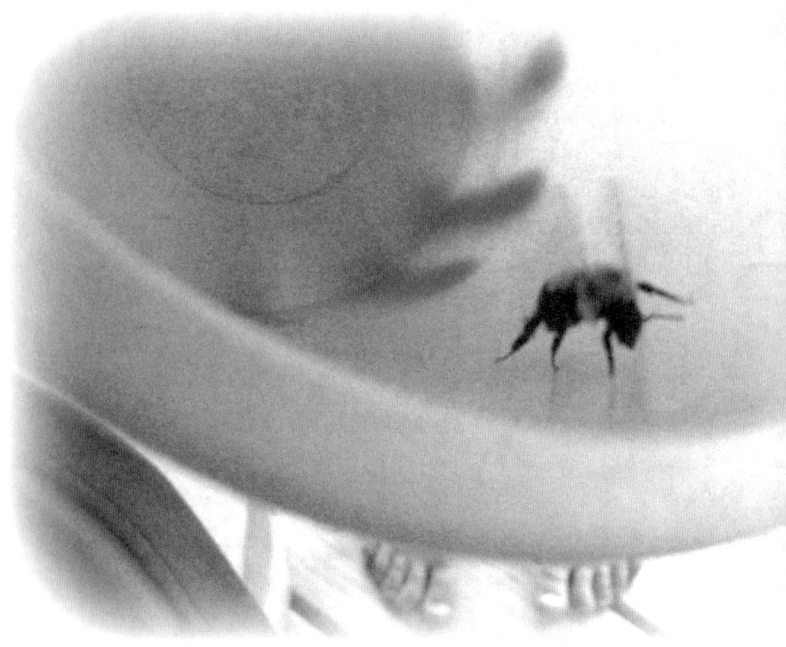

Photo by author

What Not to Do as the World Burns

> The humanities have given consistent intellectual support to the environmental exploitation which is the most distinctive product of Western civilization, and they began their work centuries before engineers became clever enough to think up ways to implement their ideas.
> —Joseph Meeker, *The Comedy of Survival*

> Your University professors are bound to preach optimism; and it is an agreeable and easy task to upset their theories.
> —Arthur Schopenhauer, *Studies in Pessimism*

Literary and cultural critics, philosophers, historians, classicists, humanities scholars in general: we keep to our side of the fence. It's safer over here, the grass is greener, the conversation is a little more, shall we say, civilized? Very few chemists or mathematicians ever wander through the gate and start shouting orders. And it's probably not wrong to say that social scientists believe—ok, they *know*—they've better things to do. On the other hand, a good many curious humanists—and let's acknowledge that not all humanities scholars are *humanists* (that is, subscribers to the doctrine of humanism, which gives prime importance to the value and significance of humans), but let's also acknowledge that most of them are—a good many curious humanists peek over the fence to see what's happening in the other paddock and, more embarrassing to the rest of us, sometimes will start aping the lingo of the physicist or the cognitive biologist. ("Art is no less a function of genetic factors than pendulous earlobes or hair on the second knuckle.")

What keeps the average humanist confined to her pen? Why, the call of the human, of course! Humanists study this fascinating species in all its majesty—what it does, what it says, the best and worst of what it's dreamed and built. There's plenty to ponder: art, story, song, cultural productions of all sorts; a bewildering variety of religions, beliefs, fancies, and penchants; plethoras of languages, symbolic forms, semiotic systems; wars, peaces, migrations, politics. Humanists are the people fit to discuss something we call "the human condition," and that condition is a work in progress, though the old questions always apply: who are we, where are we going, what should we do? These questions must be asked anew each moment as we march (or lurch) toward our brave new uncertain future, because the human condition won't lie still, and the answers you thought you had yesterday slipped through your fingers today. So lots of scholarly fodder here, and always more on the way.

Always more. But even the most eagled-eyed humanist seems to exclude from his critical gaze a rather significant fact—a rather alarming, massively unpleasant fact—which is that this focal species, this pearl of creation, has run its course. That's right: *Homo sapiens* has reached the cliff's edge and may indeed have passed beyond it, still blithely unaware, like Wile E. Coyote, that its legs are churning vapor. I'm not talking about Victorian fears of the heat death of the universe, Spenglerian pessimism based on the metaphysics of cyclical history, or even Orwellian visions of creeping totalitarianism. I simply mean that there is overwhelming scientific evidence that the ecosphere, ridden too hard for too long by too many humans, is buckling under our weight. We cannot say that a collapse will be avoided. Indications are that it won't.

In fact, as I argued earlier, the collapse has been occurring for a very long time; it's simply that now it is unfolding at a rate that makes the usual denials unconvincing. You know the facts: the last 10,000 cheetahs, 1,000 rhinos, 100 cranes; the end of pollinators and bats and birdsong; the earth fracked and flayed as if the Land of Mordor. You can't have missed the gouges, the scrapes, the permanent lesions. The tumbled mountains. The shorn forests. The buried streams. Each year hotter than the last. Floods, hurricanes, droughts, and tornadoes: more frequent, more intense. The arctic death spiral. We learn of the liquefaction of the cryosphere, the acidification of the sea, the soybeanization of the rain forest, the carbonization of the sky. All in all, you have to agree, a bungled stewardship. But humanists turn away from this unpleasantness, as if their work shall have no truck with the dying earth. Their study of the human condition scrupulously avoids the true condition of possibility for human excellence: the degradation and erasure of all things nonhuman.

Ironically, doom and gloom have always been prominent themes in the materials they study. (If only they would stay themes!) For example, one of the most obvious trends in contemporary cinema is the upswing in storylines propelled by environmental dread. You've seen these films: *The Day after Tomorrow, The Road, Children of Men, Contagion, Take Shelter, Beasts of the Southern Wild, Interstellar, Snowpiercer.* The list is long and gets longer as we push deeper into this gut-wrenching century.

Humanists distance themselves from these creative renderings of our extinctionist regime, treating them as if they were just more entries in the ledger of finally harmless speculative film. Such work, they seem to suggest, butts up against the Real in the way all art does, which is to say, only in the abstract. In art is truth, but not the kind of truth you can set your watch by.

It's the kind of truth Plato liked—and Jesus, and Emerson, and the Hallmark Card company, the kind of truth that is located safely *above* the ground, 50,000 feet in the air, the kind of truth you pass on the way to Heaven or Mount Olympus. So humanists put these horror-invoking artifacts in the same box with the rest of the poignant narratives of our earthly life: "Yes, yes, it's a rich tapestry. All the colorful threads of the human experience command our attention."

But to routinize these declinist narratives as if they were little more than twenty-first-century versions of the mutant monster drive-in flicks of the H-bomb era is to misread their significance: Ours is a time when the threats are less definable (that is, no sword of mutually assured destruction dangling over our heads) yet actually more substantive. It's not a question of *what might happen* but of *what is happening right now*. Unlike the awesome nuclear blast that destroys in a few seconds almost everything and then the rest over time as radiation shreds flesh and genomes, eco-bombs reverse the horror. They are imperceptible at first, instead gathering and multiplying their killing power over time. In other words, the horror is already on us long before we wake up to it. The genres of contemporary eco-dread proceed with the same frog-in-a-pot approach: By the time the disaster heats up it's much too late to do anything. The films don't begin with giant ants or 50-meter reptiles that in the end are brought down by intrepid reporters or cool-headed scientists. Instead eco-dread patiently assembles the signs of impending natural disaster or bodily contamination, doling out the fear bit by bit until the terrible moment when the immense wave appears on the horizon or the virus jumps the species barrier. Then all bets are off: horror touches everything. The mood here is dark and fatalistic, the concept of "the future" under erasure. The atmosphere of doom lingers over the credits as the pulsing score plays on, urging the nerveless audience to the exits. Eco-dread promises no overcoming, no Big Science or Big Military riding to the rescue. The protagonists may survive, but unlike the 1950s chillers, these movies don't assure us that civilization will.

I don't understand. Why aren't more humanists tearing out their hair over the passing of their subject, as if their own lives hung in the balance, which maybe they do? Could the answer be that ecocide is the business of those folks on the other side of the fence ("Sure, I know the oceans are dying but my gig's Shakespeare"). Or maybe it's because this niggling detail, The Crash of Man, if considered too closely, overshadows The Triumph of Man, which is the controlling motif, as though by institutional fiat, of every humanist analysis of every human production. Therefore, a constant forgetting, a convenient blind spot that occludes the signs of extinction, must be cultivated to keep the talking points on the Human Pageant forever fresh and spritely. What are humanists, after all, if not the tribunes of culture? And who needs them if they can't stick to the script? Not only are they ill-disposed to talk about collapse but they also have been superbly trained not to. Like efficiency experts at the bomb factory, they keep their heads down, their noses in the details, and their minds off the obscene end products.

But maybe it's because before they are scholars, humanities professors are people. And people don't like bad news, especially news that tells them that their entire way of life, which they have always known and on which they have built every joy and based every hope, is an ongoing calamity. That all along, below the shiny surfaces and gleaming tomorrows were the rotten pilings and sucking black mire. That this killing way of life must be stopped if life is to go on. That most likely it cannot be stopped. That we committed ourselves to collapse long ago.

News like this is hard to handle; nobody knows quite what to do with it. It's like the corrosive blood of the creature in *Alien*, eating through everything. So don't poke this beast. Keep the hatches closed. Let it rampage in the dark, down in the lower decks. Like most everyone else, humanists would prefer to speak of pleasant things and remain silent about the unthinkable. But by turning away, scholars in the humanities shirk a responsibility to bring their particular insights to the pressing problems—and rapidly closing windows for change—that lie before us. Worse, they sound like the officious Burke character in the *Alien* sequel, who, after witnessing the creatures slaughter most of the crew, nevertheless prefers a wait-and-see approach: "I know this is an emotional moment for all of us, okay? I know that. But let's not make snap judgments, please."

Tom Cohen also notices this curious silence on the part of those critical intellects charged with noticing *what is going on*:

> If a double logic of *eco-eco* disaster overlaps with the epoch in deep time geologists now refer to as the "anthropocene," what critical re-orientations, today, contest what has been characterized as a collective blind or psychotic foreclosure? Nor can one place the blame at the feet alone of an accidental and evil "1%" of corporate culture alone, since an old style revolutionary model does not emerge from this exitless network of systems. More interesting is the way that "theory," with its nostalgic agendas for a properly political world of genuine praxis or feeling has been complicit in its fashion. How might one read the implicit, unseen collaboration that critical agendas coming out of twentieth century master-texts unwittingly maintained with the accelerated trajectories in question? The mesmerizing fixation with cultural histories, the ethics of "others,"

the enhancement of subjectivities, "human rights" and institutions of power not only partook of this occlusion but 'we theorists' have deferred addressing biospheric collapse, mass extinction events, or the implications of resource wars and "population" culling. It is our sense of justified propriety—our defense of cultures, affects, bodies and others—that allows us to remain secure in our homeland, unaware of all the ruses that maintain that spurious home. (15)

I won't presume to tell anyone how to approach the human condition in the age of environmental catastrophe. Part of the charm of humanistic inquiry is that it has as many directions as there are scholars. But it does seem to me that the questions Al Gore posed in his Nobel acceptance speech are apt framing principles for our disciplines and the kind of education we are providing in this, the century of environmental reckoning:

> The future is knocking at our door right now. Make no mistake, the next generation will ask us one of two questions. Either they will ask: "What were you thinking; why didn't you act?" Or they will ask instead: "How did you find the moral courage to rise and successfully resolve a crisis that so many said was impossible to solve?"

Clearly, the humanities—arguably, more than any part of the university—have a special relationship with the "next generation." Humanists are very good at summoning history in order to give context to the present and, sometimes, at projectively rendering the future based on history's precedents. And as we task our students to contemplate the lessons of the past, we are tacitly pushing them from their comfort zones in the seemingly stable now and prompting them to spin out answers to the same question John W. Campbell, the great editor of the Golden Age of science fiction, claimed was central to his genre's mandate: "If this goes on . . . ?"

In other words, humanities people have long been uncomfortable futurists. Yet one of the unstated requisites for becoming a scholar of the humanities is that on a professional level you must give up the notion that you are working in a biosphere. You come to believe, quite organically I think, that because you are for the most part producing texts about other texts, the orders of meaning and representation take precedence over the orders of matter and energy. You tend to imagine, for example, that nightingales in Keats or albatrosses in Coleridge are more substantial than the noisy birds at your feeder. (They are possibly starlings, by the way, introduced by literary societies into North America in the nineteenth century to ensure that the New World would boast all the Old World birds that Shakespeare named.) These textual entities are the humanist's bread and butter, and though the real ones are all well and good, they don't figure much in the humanist's methods and materials. They and their forerunners have always worked at some remove from the world of chalk and cheese. Long before the advent of the digital computer, philologists were immersing themselves in the hot bath of information. In fact, nothing much has changed since the time of the Babylonian scribes. Signs took on powers and dominions beyond the mundane world of dirt and wind. Could dirt sing the deeds of god and heroes? Could the wind style the lineage of kings or number the bushels of wheat in the granaries? No. Words and pictures, representations of all kinds—these attracted the attention of a certain type of thinker, and as a result magnificently potent, abstract modes of cognition emerged. Aesthetics, rhetoric, history, metaphysics—all rely on a coding of material existence into systems of signs, which can then be rearticulated in ways no longer beholden to any objective reality. But this type of knowing—speculative, inferential, introspective—always runs the risk of the echo chamber effect, whereby texts come to talk to texts alone, never getting outside the room for a breath of fresh air.

If I'm being too elliptical, let me offer an example. Not long ago I received a call for papers from an environmental humanities conference in Germany. Ostensibly, the conference should be a dream come true for an antihumanist like me: "We are looking for contributors to a transdisciplinary symposium on the didactical implementations of ecocriticism, critical animal studies, and green cultural studies. With a special emphasis on transdisciplinary perspectives, we would like to discuss how the tenets of these academic fields can be incorporated into the daily practice of teaching the humanities and arts—" So far so good. But here is the other side of that dash that sends this symposium off into the bubble of humanist boosterism: "—without either breaching the topics' complexity, falling into the mode of environmentalist propaganda or succumbing to warnings and claims to catastrophic urgency which are hard to reconcile with an ethos of critical and democratic pedagogy." In other words, the very thing that in the last thirty years has prompted many humanists to question humanism—that is, the "catastrophic urgency" of our environmental moment—is dismissed out of hand, as if serious scholars shall have no truck or trade with planetary catastrophe in its full horror. To succumb to any such urgencies would be to let slip, evidently, the reasoning mask so carefully secured over many centuries, to allow the raucous street fight to disturb the high-toned discussion going on in the parlor.

A caveat or two is in order. There is a big difference between my critique and what some, following in the tradition of C. P. Snow, have to say about the willful dismissal of scientific facts perpetrated by humanists. The science writer Timothy Ferris, for example, notes,

> Being an intellectual [read humanist] had more to do with fashioning fresh ideas than with finding fresh facts. Facts used to be scarce on the ground anyway, so it was easy to skirt or ignore them while constructing an argument. The wildly popular 18th-century thinker Jean-Jacques Rousseau, whose disciples range from Robespierre and Hitler to the anti-vaccination crusaders currently bringing San Francisco to the brink of a public health crisis, built an entire philosophy (nature good, civilization bad) on almost no facts at all. Karl Marx studiously ignored the improving living standards of working-class Londoners—he visited no factories and interviewed not a single worker—while writing *Das Kapital*, which declared it an "iron law" that the lot of the proletariat must be getting worse. The 20th-century philosopher of science Paul Feyerabend boasted of having lectured on cosmology "without mentioning a single fact." Eventually it became fashionable in intellectual circles to assert that there was no such thing as a fact, or at least not an objective fact. Instead, many intellectuals maintained, facts depend on the perspective from which they are adduced. Millions were taught as much in schools; many still believe it today.

While I understand the motive behind it, Ferris's truculent view is mistaken for a number of reasons. Suffice it to say that if he believes only humanist intellectuals produce factitious realities, he has little understanding of what ideology is. His discussion is actually rife with ideologically charged simplifications, propelled, no doubt, by his unblinking scientific realism. What could be more ideologically saturated than Ferris's proposition that only those immersed in objective facts can proffer an ideology-free position? That's like saying only polar bears have any business talking about snow.

A better way to describe the two cultures divide is to note that while scientists show great and continuing interest in objective reality, humanists are inclined to stipulate to it and move on. They know it's there, they know it's important, but mediating between objective reality and their work product is not in their job description. Their job actually begins where the trail of objective reality peters out. The so-called ideology that, according to Ferris an inattention to facts produces, is actually what humanists are trying to track: let's call this place, this asylum from objective facts, "human culture." Here facts are much more difficult to nail down than in the lab: there are lots of facts, they don't wear name tags, and they *necessarily* look different depending on where you stand. For me, it's not getting things wrong that's the danger; where *right* is a moving target anyway, being *wrong* is just an occupational hazard.

It's rather the myopic focus on the system of culture that I find dicey. Given the unfolding biospheric crisis, humanists clearly need to produce more humane understanding of extra-human reality. But I think we have to understand a bit about the history and current trajectory of the humanities disciplines to see why they are so unresponsive to environmental exigencies. From Johann Herder and Matthew Arnold, through Wilhelm Dilthey and Hans-Georg Gadamer, to present-day cultural scholars like Harold Bloom and Camille Paglia, we find a construction of humanities as *the* essential disciplines that must foster a nonutilitarian perspective on human life and purpose— never more important than now, in a world become so fixated on bottom-line deliverables. In the glare of naked economism, the humanities' inquiry into nonutilitarian values does itself look to lack utility—hence the humanities' attempts to up their value by touting their capacity to teach soft skills and by creating alliances with other, more "useful" branches of learning such as business or computer science.

The danger, clear and present, for humanities disciplines embracing these alliances is that they may begin to believe exactly what they need to contest: the notion that all learning ultimately needs to prove itself in the market. Yet we know the market is a blind worm. As it aimlessly churns its way about the earth, its only goal is to process what it finds through its code of profit and loss. And, of course, most of what ails this planet is either produced by our commitment to that coding or by our failure to attend adequately to what the code leaves out. Those aligned with the code self-blind themselves eagerly and without remorse. "You can't make a man understand something when his salary depends on him not understanding it," said Upton Sinclair. Or, even more apropos of this discussion, as the arch-climate-change denier and sometime chair of the Senate Environment Committee James Inhofe said to Rachel Maddow, "I was actually on your side of this issue when I was chairing that committee and I first heard about this [global warming]. I thought it must be true until I found out what it cost" (Maddow). We in the humanities must avoid being co-opted by calls for utility, no matter what pressures are exerted.

This situation pertains as much to universities as to individual disciplines. David Ehrenfeld, conservation biologist and conscientious objector to the arrogance of humanism, makes the following observation, worth noting in full:

> The problems that the universities are doing little or nothing to address—either in teaching or research—are those

that we must confront if our civilization is to survive. They are the materialism in our culture; the deterioration of human communities; anomie; the commercialization (privatization) of former communal functions such as health, charity, and communication; the growth imperative; exploitation of the Third World; the disintegration of agriculture; our ignorance of the ecology of disease, especially epidemic disease; the loss of important skills and knowledge; the devastating decline in the moral and cultural-intellectual education of children; the impoverishment and devaluation of language; and the turning away from environmental and human realities in favor of thin, life-sucking electronic substitutes. Far from confronting these problems, universities are increasingly allying themselves with the multinational commercial forces that are causing them. The institutions that are supposed to be generating the ideas that nourish and sustain society have abandoned this function in their quest for cash. (188–89)

There are at least two implications we should draw from this passage. The first is that the true audience of the university is exactly the audience few academicians really want to imagine they're playing to: the nakedly commercial interests that pick up our ideas and repurpose them (and our students) for financial gain in the real world. This is a paying audience; it ponies up its own dough in the form of research and consultancy contracts, co-op and summer jobs for students, and high-profile donations and endowments. But in point of fact, the private sector's tab is mostly picked up by the public, under the quasi-logical assumption that they will be repaid in the form of employment and economic growth. This apparently virtuous circle takes on the look of a noose as the marketplace tightens its control on academic priorities through a range of entanglements. The university's very self-description, for example, has been colonized by the language of commerce. This remetaphorization figures the university as corporation, the student as client, and the professor as manager. If professors are only now getting used to this discursive shift, the university's administrators have already embraced the reality. How else to explain the explosion of liberal arts and sciences programs that marry business courses to discipline X, or the preponderance of grants and funding opportunities that openly direct research toward the purported needs of various sectors of the economy? (I laugh now at my own naïveté when, as a young assistant professor chairing a PhD defense in engineering, I was shocked to hear the candidate in his preamble thank his corporate partners and sponsors, complete with PowerPoint slides bearing their corporate logos.)

How else but by the triumph of private over intellectual enterprise to explain the rhetoric of excellence, innovation, and commercialization that issues so steadily from the public faces of the academy? This rhetoric, I'm afraid, is not empty rhetoric; it is rather the clear articulation of the belief system that animates the university at its highest levels. I vividly remember the then chancellor of my university—and head honcho of RIM-Blackberry—speaking to the graduating class of the arts faculty a few years ago: "When I look at you, I don't see a room full of graduands but rather 3,000 technology transfers." I scratched my head in puzzlement, as did any number of the assembled, pondering the high-tech potential of, say, a classical studies major. But the valediction was in keeping with his general view of the role of universities as repositories and manufactories of knowledge, expertise, and ingenuity, all of which are to be piped seamlessly into the wealth-generating engines of the nation. I was bowled over, but not so much that I failed to transcribe more of his remarks:

> If you really want to understand commercialization, all you have to do is attend convocation at your local university. At mine, the University of Waterloo, we celebrate—yes celebrate—the passage of the next generation of students into the economy and society twice each year. Armed with cutting-edge technology from around the world, the latest tools, the latest techniques and processes learned from their work under the very best researchers, they graduate with much fanfare and go on to build the industry, institutions, and society of our country. Now that is real commercialization.

I suppose this sort of guff is meant to be cheering, even uplifting, to bright-eyed graduates about to tackle an uncertain job market—and to us eggheads left behind in the ivory tower and already looking ahead to the next batch of trainees.

But the narrative has some troubling loose ends: What becomes of knowledge that can't be commercialized? What of knowledge that asserts that the quest for commercialization is precisely what undermines a society's values and institutions, not to mention its ecological bases? What of knowledge that says, heretically, that we simply shouldn't build the next money-making gadget because, frankly, it makes the world worse? The drawback of the-university-is-a-knowledge-factory metaphor is that it casts professors as the parts suppliers and students as the parts—no matter that the machine they're about to be bolted onto to is a dangerous piece of junk. The age-old critical and speculative functions of the university are set aside, and free inquiry into the problems, mysteries, and injustices of our world is replaced by motivated research into what sells. It becomes difficult to incorporate perspectives that run against this conventional wisdom.

If there are still such animals as "curiosity-driven research" or the "pursuit of knowledge for its own sake," it's a fair bet that before much longer they will be found padding around in only the most obscure and underfunded corners of the university. In the modern university, faculties eat what they kill, and if your teaching and research cannot project itself "beyond academia" (read, "to the marketplace"), it will soon find itself at the bottom of the food chain. So the pressure mounts to fully integrate academia into the power-knowledge-cash nexus that already insinuates itself, virus-like, into most earthly considerations outside the university. If they cannot come up with a viable alternative mission, very few professors in the years to come will be able to conduct what was once quaintly labeled *pure research* and *teaching*. With no shelter from the storms of progress, every scholar will have to keep a weather eye out for the winds that will either fill her sails or swamp her.

The second implication of Eherenfeld's observation is the one that I think more directly pertinent to my discussion: that humanities departments have an absolutely crucial role in keeping alive—or at least groping toward—that alternative mission. If the majority of disciplinary formations are increasingly forced to justify themselves on the basis of their relevance to the trends, requirements, and ambitions of contemporary disaster capitalism, the countervailing impulse of humanities departments is to imagine modes of life on Earth not rooted in utility and quarterly profit reports.

I don't deny that much work must be done to green the humanities. These disciplines are constitutionally anthropocentric in that the matters they study must always return to the tragedies and triumphs of man [*sic*], the complexities of his social and cultural networks, the challenges of his self-created media and technological cocoons, the booms and busts of his business cycles, and the pathologies of his legal, political, and spiritual orders. In short, man is the fixed pole, the measure of all things, and there's not much chance that *les sciences humaines* are ever going to look away from the human drama, next to which the biophysical world is simply stage dressing. Now, as they accumulate the trappings and ethos of professional schools, the humanistic disciplines have become even more invested in the debilitating fantasy that nothing lies beyond the reach and purview of man. And this at a time when what we really need is a massive surge of humility. I quote the distinguished Oberlin professor David Orr to give you a flavor of what we are up against:

> We are still educating the young as if there were no planetary emergency. . . . The crisis we face is first and foremost one of mind, perception, and values; hence it is a challenge to those institutions presuming to shape minds, perceptions, and values. It is an educational challenge. More of the same kind of education can only make things worse. (27)

Needless to say, the roadblocks to incorporating an environmental perspective in the humanities are formidable. I have spent much time explaining why humanists are ill-disposed to talk about these matters and how the market-driven principles that are colonizing the humanities make doing so even harder. At this point in my discussion, readers will expect me to outline how the humanities can help blunt the onrush of ecological calamity. What should educators and students do to make a significant and durable contribution on the right side of history? It's time for the obligatory note of hope and a series of action steps.

I hate to disappoint. But I'd rather conclude in the mold of "disillusioned cynic" that I've constructed in this essay. A small anecdote provides some partial context. Some years ago, our faculty wished to set up a new liberal arts satellite campus in a nearby town. An open call was sent out to all of us: dream big, nothing off the table, blue-sky proposals invited! A meeting was scheduled. Two professors came forward with proposals; I was one of them. Rather than a curriculum pegged to a common humanities core, as has often been the case in liberal arts colleges, my proposal described a new core of humane studies rooted in principles of sustainability and ecological holism. In this core, all the courses, texts, and modes of delivery would be held to a single standard: How do they help students discuss, imagine, and prefigure an environmentally sound future?

Unlike Ehrenfeld, I've never thought the humanities have had a whole lot to tell environmental studies; rather, I think it's the humanities that are in dire need of environmental education. What the human sciences require, therefore, is more and frequent encounters with extrahuman reality; they could stand a little "de-humanization." At the risk of again sounding naïve, I will quote the heart of my proposal.

> I believe our goal should be something more radical and yet more truly conservative than the creation of another small college replicating a liberal humanist model of anthropocentric learning. That sort of college is geared toward students who wish to follow individual routes to individual careers and individual rewards. I propose instead that we create a college that will shape minds, perceptions, and values in ways that take account of the true price we are exacting from this planet. I propose that we seek to draw a different kind of student and produce a different kind of graduate. What kind of graduate? Well, our world needs economists who understand no cost is ever externalized; it needs entrepreneurs who create wealth without endangering the planet's health; it needs politicians who measure policies in terms of their effects on people as yet

unborn; it needs writers and artists who appreciate Earth itself as the greatest work of all; and it needs teachers who will communicate the message of sustainability to children in the critical moments when they are just beginning to form their own values. In short, our world needs graduates who can help save it by foregrounding the environmental problematic in every walk of life—which is exactly where liberal arts graduates wind up. There are no higher stakes, and I hope we could consider the new venture as one very significant means to this noble end.

The dean seemed pleased with my lofty rhetoric, but the proposal never went further than its initial presentation. He and I both knew that the university, the various levels of government, and the private partners who were lining up to support the new college would have no interest in undermining through education the promarket paradigm to which they were beholden. He and I both knew that the purpose of education is no longer—if it ever was—to challenge the assumptions of the financial, political, and technological orders that have produced a decaying society and a dying planet. He and I both knew that the purpose of higher education is to deliver on time those job-ready graduates that our corporate friends had already put in orders for, and to serve as tax-subsidized research incubators for product development.

During the meeting, another possible focus of the new campus was hinted at: digital media and global business. Although no one had stepped forward to present on these areas, I realized immediately that here had been the real agenda all along. Two areas, one might say, that market wisdom was suggesting could provide a strong return on investment.

How did this story end? As the satellite campus initiative

gained momentum, new entailments were grafted on as others were lopped off (including the whole "liberal arts" concept). According to information disseminated at the time, the new institute would focus on digital *convergence* and *aggregation*. This capacity to bring together key *partners* would be the institute's *core value proposition*. The campus would have undergraduate and graduate courses that featured *client-driven assignments*. Its mandate read a lot like the chancellor's vision of the ideal university: a place where "students, leading researchers, businesses and entrepreneurs [come] together to create, examine and commercialize opportunities in the digital media space." Notice that the objects of study were framed not as problems, gaps, lines of inquiries, anomalies, or enigmas but rather as "opportunities," which is yet another term imported from nonacademic discourse. The term implies that only research that promises commercial return is of interest. Outside-the-box thinking will be harnessed for inside-the-box ends. Compliant creativity.

I'll conclude with a blunt assessment. If we take this new campus as a model for the university to come, we are in deep trouble. This model does not possess the moral vision to go after the hard problems "we must confront if civilization is to survive," for it has been captured and domesticated by moneyed interests. Thus it cannot entertain alternative views of its own future because its fortunes depend on the deliberate unknowing of its history and the reproduction of its irresponsible present. Montaigne wrote, "Ignorance that knows itself, that judges itself and condemns itself, is not complete ignorance: to be that, it must be ignorant of itself." Sadly, the modern university is rapidly moving to replace productive ignorance with plain ignorance through willful self-blinding.

Its perspective does recall, however, the old Chinese motto:

crisis is opportunity. One can imagine this kind of university, like any shrewd business, making lemonade if the world hands it lemons; it might actually turn a profit in a planetary emergency merely by going about its task of cranking out new techno-fixes and pleasing distractions as it pursues its visually impaired and ultimately self-defeating autopoiesis. But when the extravagant dreams of today's leaders of the university are dashed tomorrow on the rocks of climate change, species loss, resource scarcity, and so on, let us hope that this will not have been because the *opportunity* wasn't also understood, at least somewhere in the university, in some beleaguered set of disciplines, as exactly what it is: a *crisis*.

Photo by author

The Denialists, or What If This Present Were the World's Last Night?

> The basis of optimism is sheer terror.
> —Oscar Wilde, *The Picture of Dorian Gray*

CIVILIZATION ANYWAY

With apologies to Walter Benjamin and Paul Klee, the angel of history has been grounded. He teeters atop a smoking pile of trash, gazing out over a vast pit. The pit was formed during the brief period when matter and energy could be drawn cheaply from the earth and used to build and power our great industrial civilization—this lunatic machine with no off switch that is launching us face first toward a brick wall. The angel is famously backward looking, but if he could crane his head around to the future he would see nothing but wreckage to the horizon, a salted landscape under a bleeding sky.

We would like to get a handle on this dirty business of ecocide, geocide, planetcide. We seek a name, a frame, a scheme that could encompass the worst of us. Capitalism, patriarchy, logocentrism, speciesism, Modernity. Nothing quite captures it. The river of excess is too wide. We cannot bridge it, for to do so presumes there is another side. But there is none. It is an ocean-river, a dirty flood that carries with it everything in sight. And there is not even a near shore from which to view it. I myself do not stand apart from the deluge. I bob among the chicken nuggets and curly fries, the unexploded mines and disposable razors, the capsized luxury ships and one-click order buttons. Let's call it what it is: civilization, the whole roiling bulk of it, from Uruk to New York City.

What is civilization anyway? The British historian Felipe Fernández-Armesto understands civilization to mean "a relationship between man and nature" (30). This definition avoids the messy challenge of toting up the putative features of a model civilization (for example, urbanization, state institutions, militaries, communication and transportation networks, cultural productions) and then applying them to other target societies, each of which undoubtedly violates the lading list in some respect and therefore is at least partly uncivilized. Instead, he prefers to investigate the extent to which different societies have been able to reconfigure natural systems for their own purposes. Societies that have altered their environments in very profound ways fall on one end of the civilization continuum, while those less deft at resource exploitation and development fall at the other. By Fernández-Armesto's account, today's planetary ecumene is, no surprise here, by far the most civilized in history. "More civilized" does not mean better, he cautions, but he does feel a hearty admiration for the distance we've come:

> The world is a place of experiment—an expendable speck in a vast cosmos. It is too durable to perish because of us. But it will surely perish anyway. Our own occupancy of it is a short-term tenancy.... We ought to make the most of it while we have it. This may be more satisfactorily achieved by a sort of cosmic binge—a daring self-indulgence of the urge to civilize—than by a prudent and conservative desire to protract our own history. Just as I would rather live strenuously and die soon than fester indefinitely in inert contentment, so I would rather belong to a civilization which changes the world, at the risk of self-immolation, than to a modestly "sustainable" society. (34)

A bracing vision, to be sure. The Kalahari Bushmen have run a steady state operation for the last 20,000 years, but oh, the wonders of the Burj Khalifa soaring above the sands of Dubai! Will you settle for rock paintings and bush meat when you could sip champagne on the 124th floor with a view to the artificial pleasure islets in the Persian Gulf? If you set your sights on your ankles, life looks like a dreary trudge through the dust. Just admit it: We're here for a good time, not a long time. The Aztecs feted their god-impersonator for a year before they cut out his heart and ate his flesh: we are doing the same thing, just on a broader scale and with ourselves as both priest and sacrifice.

Perhaps that's a bit unfair to the Aztecs. After all, their ritual killings were meant to appease and honor the gods, whereas our gluttonous self-murder is meant to propitiate only ourselves. Still, if you find Fernández-Armesto's devil-may-care attitude appealing, you may well be on the right side of history. The last few decades have proven there is little reward for taking the long view; indeed, all good things come to those who don't wait. The moment of opportunity is now. Carpe diem. Go for it! YOLO. We postmoderns cannot conceive of the future as the offspring of today; tomorrow is the child of another mother. Its occupants—even if they are our older selves—are no kin to us. Beyond wishing them Godspeed, we feel we owe them very little. I can no more commiserate with my future self of a year or a decade hence than I can commune with the dead. Above my walkway spreads a fast-growing poplar: why should I plant an oak that will give shade only to my grandchildren? Posterity ain't what it used to be.

In his *Reason and Persons*, the philosopher Derek Parfit suggests this disinclination to give thought to the future stems from the basic psychological orientation of individuals toward their time-shifted selves. Although we may feel mental and physical continuity with past versions of ourselves—nothing like memories and scars to confirm you are who you think you are!—it's much more difficult to project that continuity forward, to the as-yet-to-be you. Parfit argues that what we imagine as a thinking self, a *cogito*, to borrow Descartes's term, is more "like a nation" (275). Personal identity doesn't count for much in this reductionist view; rather, what counts are the states and events in our brains that contribute to our narrative of psychological unity. The significance here for individuals—and by extension civilizations made up of these short-sighted beings—is that our empathy for future iterations of our selves is hardwired to be very low, perhaps no greater than for strangers. It shouldn't surprise us that we do all sorts of things that we'll regret years down the road: take up smoking, overeat, cover ourselves in garish tattoos, and so on. Like any good corporation, we have an eye on the quarterly profit statement, not the ten-year forecast. Our future selves—our future societies—are thus deeply discounted, so much so that some thought-leaders suggest sloughing off the environmental problems of today onto the supposedly richer and smarter generations to follow. Just as my older, wiser self will be better equipped to work off the flab I have accumulated over the last months of reckless self-indulgence, future societies will possess the superior wealth and sterner politics needed to handle the thorny environmental problems we don't want to face today. If they fail to do so, well, that, as Fernández-Armesto might say, is the risk of self-immolation that any civilization runs.

OUR ASTEROID PROBLEM

But perhaps you don't want to assume the risk. Perhaps you'd feel better knowing the future was secure, if a bit hazy in its details. Perhaps you're one of those softies for whom self-immolation just doesn't appeal. You are not alone. In *Death and the Afterlife*, Samuel Scheffler argues that a large part of what gives individuals the courage to continue living with the certain knowledge of their own inevitable extinction is that they are equally confident the world itself will carry on after they're gone. With this sort of "afterlife" assured, one's efforts during one's lifetime can be imagined to accrue to a grander trajectory of human presence and purpose. To arrive at that conclusion, Scheffler imagines a scenario in which a person knows with absolute certainty that thirty days after he dies the planet will be destroyed in a collision with an asteroid. His life will be perfectly normal up until his death, with all avenues for happiness and satisfaction remaining open, but it will have to be undertaken with the clear foreknowledge that nothing he knows and loves will remain one month after he departs. As Scheffler points out, this scenario bears some resemblance to the sterile future history of P. D. James's *Children of Men*, in which a fertility crisis has ensured that no children will ever again be born, so that human life will disappear when the last man or woman succumbs to old age. In the novel James also explores the emotional, cultural, and spiritual dynamics of a society ticking down to termination. Drawing on such scenarios, Scheffler deduces, convincingly, that the prospect of our species' extinction is actually far more detrimental to psychological well-being than the prospect of one's own death. In other words, that one can take for granted a viable collective future underwrites one's sense of personal significance. Without that confidence, individual equanimity becomes near impossible. Scheffler finds the futureless world to be "characterized by widespread apathy, anomie, and despair; by the erosion of social institutions and social solidarity; by the deterioration of the physical environment; and by a pervasive loss of conviction about the value or point of many activities" (40).

In his view, the presumption of collective survival is the condition of possibility for our sense of the importance of our present lives and our contributions to the ongoing human project. "To put it a bit too simply: what is necessary to sustain our confidence in our values is that we should die and that others should live" (108). This is so because, for example, many of the ongoing activities we are engaged in—from cancer research to space exploration to simply providing a nest egg for our children or their children—will not come to fruition until after we are gone, and thus it's precisely the prospect of this afterlife for our efforts in the present that makes life satisfying despite its relative brevity. Scheffler's conjectures, measured on the pulse, feel resoundingly true. In Marcel Theroux's novel *Far North*, one of the first true climate change fictions, the protagonist Makepeace comes to a similar conclusion about the value of existence in the face of annihilation:

> A sane person knows they're headed for the end of something. But the thought that things will continue, that there'll be kind words at their funeral, or even just the pulse of blood in someone, somewhere, that dumbly recall that they were here—that gives the rest of it some point. A sane person expects that. . . . Everyone expects to be at the end of something. What no one expects is to be at the end of *everything*. (241–42)

Parfit's perspective seems on first blush to be incompatible with Scheffler's. In essence, the former says that the social order's survival (that is, the continuity of civilization) is a foreign country; the latter says we cherish that country as if it were our own. But if we view Parfit's future-heedless individual through the lens of Scheffler, we discover this person can afford to be so blithe about tomorrow only because he cannot imagine it won't always be there. Scheffler reminds us that who we take ourselves to be and where our society believes it is headed hinge on a central—dubious—assumption: that our world, that our species, has a viable future. Like the ancient pharaohs, civilizations behave as if they are death proof, and they conscript their members to contribute unquestioningly to this conceit. (As Schiller put it, the "concrete individual life is extinguished, in order that the abstract whole may continue its miserable life" [48].) This grand self-assuredness empowers but simultaneously sedates, for while we mere persons are chuffed to feel our acts are meaningful and important, we are ill-prepared when evidence emerges that they are not. We become unwilling to consider that our quotidian works and days may not be backstopped by a larger permanence. And thus we are dismissive of the terrifying truth: the asteroid is *already* hurtling toward the earth and will very soon put an end to all our pretensions.

Helpfully, Ben Winters's *The Last Policeman* trilogy examines these very issues in the handy form of a police procedural. Set in the months and weeks before the strike of a six-kilometer-wide celestial object, detective Henry Palace—the eponymous last policeman—carries on with his investigations as if the end of the world provides no disincentive to the pursuit of truth and justice. While everyone around him is completing bucket lists, taking drugs, committing suicide, or hoarding guns and food, not Palace. He behaves as if his life—and the lives of killers, missing persons, and the innocent—are as meaningful as they ever were, asteroid or not. It isn't the case that Palace doesn't care about the coming annihilation; it's rather that he affirms the universal value of leading a dignified life, which—again, asteroid or not—will always end the same way. How to live in the here and now while under a death sentence is not a new question. As Palace says to a husband who has deserted his wife and is about to undertake a questionable act, "Respectfully, sir, the asteroid did not make you leave her. The asteroid is not making anyone do anything. It's just a big piece of rock floating through space. Anything anyone does remains their own decision" (195).

In Winters's novels, Palace is that everyman who is not actually an everyman at all. In fact, he is the everyman's antithesis, a figure going in exactly the opposite direction to that posited by Scheffler in the asteroid scenario: he doesn't believe the coming destruction of civilization annuls the laws and principles we must live by in the here and now. Kant would approve. He is that hero equipped with a binding moral code that we've been taught to admire in fiction and film. Palace: an intrepid gumshoe who perseveres and solves his cases in spite of the failing society around him. He is that Kiplingesque "Man" we'd all like to be ("If you can keep your head when all about you / Are losing theirs . . ."). In another sense, he is the rare exception that tests the Schefflerian hypothesis: an individual undaunted by the certainty of global destruction. Some readers have suggested only a person suffering from mental illness could behave as Palace does.

With even greater impact, the asteroid scenario was played out cinematically in Lars von Trier's 2011 film, *Melancholia*. Frequently viewed as a meditation on the vagaries of black depression, the film is better understood as a rehearsal of the Schefflerian afterlife hypothesis and as a parable of environmental collapse. The premise is simple: as a hitherto hidden planet in the solar system makes its way toward Earth, characters must confront the prospect of personal and species extinction. (The audience, unlike the characters, knows the collision is inevitable, for Von Trier opens the film with ten minutes of sublimely beautiful scenes of the Earth in its death throes. The emotional strain on viewers contemplating doomed characters—not an altogether unfamiliar challenge for audiences—is pushed in a new direction, for the world itself is also under a death sentence.) Some characters will simply choose to disbelieve the science. But the central character, Justine, is convinced otherwise. She is blessed or cursed with a clairvoyance that tells her not only will Earth by destroyed, but that it's a good thing too: "The earth is evil. We don't need to grieve for it."

In the first half of the film, which is set on the evening of Justine's wedding reception, she rapidly descends into a deep, angry, irresistible funk, perhaps prompted by her early presentiment of Earth's impending doom. The other characters, as yet oblivious to the incoming planet, spend much of their time dealing with the social disaster that is her behavior. In the second half of the film, Justine's earlier breakdown is somewhat beside the point, as now everyone else must face their own existential crisis. Some will side with the deniers, who are convinced that a near miss will occur; others put on a brave front only to check out ahead of time by suicide. But Justine, who gradually emerges from her postwedding state of near-catatonic depression, seems the only one equipped to face the cataclysm with any kind of composure. Her clinical condition has already prepared her for a life without hope, and as the catastrophe looms, she alone is able to endure the final, searing moments with dignity and strength.

In *Melancholia*, Von Trier suggests that there are two basic postures toward the futureless world: that of denial, which coexists with a kind of self-deluding hope, and that of acceptance, which coexists with a certain resoluteness and energy. What good this latter posture might do us is anyone's guess. If the planet is doomed, all the stiff-upper-lips in the world will go for naught. But I am reminded of Derrick Jensen's statement, "Hope is a longing for a future condition over which you have no agency. It means you are essentially powerless" (330). By contrast, when hope is gone, there is a potential to actually *do* something: "when we stop hoping for external assistance, when we stop hoping the awful situation we're in will resolve itself, when we stop hoping the situation will somehow not get worse, then we are finally free—truly free—to honestly start working to thoroughly resolve it. I would say when hope dies, action begins."

WHAT IF THIS PRESENT WERE THE WORLD'S LAST NIGHT?

One could rightly argue that in the asteroid scenario there is little point in taking action since the calamitous physics of intersecting orbits are inescapable. Why not relax and enjoy the spectacle? Indeed, this is what Justine does in *Melancholia*, basking nude in the double-shadowed night under a full moon and the bright new planet, a haunting scene of beguiling Thanatos. Meanwhile, on our ragged, fraying planet, even at three minutes to midnight, all is not yet lost. Much, but not all. For the most part, however, our species continues to behave as if the stakes still aren't high enough. We're doubling down on apocalypse. Perhaps denialism, with its hopeful mushy center, is easier to swallow than acceptance, which tasks us with that most difficult question: "So *now* what should I do?"

Cicero claimed while there is life there is hope. But, as the Mark Wahlberg character notes in *Deepwater Horizon*, "Hope ain't a tactic." And let's be clear: we are all denialists now. To take the next step beyond hope means we'd first have to pull our heads out of the sand, and so far, we are not inclined to do that. Denial goes deep into the human animal, right to the quick.

"Only man will deny the known truth" (Jeffers 411). Lionel Tiger believes that denial is bred in the bone, laid down and locked in by evolution itself, such that "neither the consciousness of mortality nor a cold sense of human frailty depresses the belief in desirable futures" (16). We hew to a willful, blind optimism that, to be fair, stands us in good stead against shorter-term nuisances like charging bulls and lost car keys. We have faith that we can move ourselves away from danger and recover the objects of desire. But the drawback is that our ingrained predilection for looking on the bright side works against us when the dark predominates; we fixate on the tiny circle of light at the end of tunnel, only dimly aware that we are in the black heart of a mountain, and the rock is beginning to shake.

Denial is an offshoot of our narcissism, itself a reaction to our fear of death. The denial of death and the self-esteem culture go hand in hand. It's no wonder we can't face up to the facts, the patterns, the science, that lay out our peril with great precision. To face those facts would mean puncturing the veneer of cosmic centrality that is the birthright of each Western woman and man. Ernest Becker writes,

> His sense of self-worth is constituted symbolically, his cherished narcissism feeds on symbols, on an abstract idea of his own worth, an idea composed of sounds, words, and images, in the air, in the mind, on paper. And this means that man's natural yearning for organismic activity, the pleasures of incorporation and expansion, can be fed limitlessly in the domain of symbols and so on into immortality. The single organism can expand into dimensions of worlds and times without moving a physical limb; it can take eternity into itself even as it gaspingly dies. (3)

As with the man, so with the culture. Facing up to collapse would reveal an impermanence no self-respecting civilization wants to admit. What is the profit in disinvesting from our visions of plenty? Like a cat batting stupidly at a televised mouse, each of us is drawn to the image of a grand tomorrow, blithely unaware that the progressive narrative is beamed into us by charlatans whose only goal is to sell us more soap.

What we always fail to grasp is the utter baselessness of our dogged optimism. We are the species that against all evidence sees our ugly present as the fertile ground for a vibrant future. But you can't polish a turd. No amount of glossy futurism can paper over the poverty of our condition. We are the doting parents with the staggeringly unhandsome child: it's natural to want to see the best in the flesh and blood you've brought forth, but at some point, you need to concede this child will never be a fashion model. Denial simply forecloses options until there is nothing left but bitter disappointment and regret. Unfortunately, denial is the way of being for the get-along guys and company men that we've all become. Each of us is committed to the fantasy that the organization is running smoothly. To step outside the holo-cave for even a moment, to gaze on the clanking machinery as it shudders and sparks, that would take courage and resolve. Better to leave the tough decisions to the experts and the managers; as for the rest of us, we'll keep our heads down. If the brain trust says we are on solid footing, then that's good enough. The holo-cave is warm and cozy; the shadows flit about invitingly.

When we do cast a glance into the middle distance, eyes squinting from years of watching our step, it's only to finger the obvious culprits and absolve everybody else. Let the blame fall on the Big Deniers! Well, they certainly do bear outsized responsibility. It's easy to inveigh against Exxon or Koch

Industries, those empires founded on the release of CO_2, or Senator Coal and Congressman Crude, who form the political arm of the oil and gas folks. It's easy to cry foul on petro-states like Canada, Norway, and Kuwait, which have their noses so deep in hydrocarbons they can see only the black. And it's easy to lay the mess at the feet of the presidents and legislators, conservative and liberal alike, who invariably elevate pocketbook issues over the more complicated matter of planetary survival. But one can spend only so much time blaming the wind for blowing.

THE HUMAN SYNDROME

No, there must be something profounder underlying all this annihilation, what's been and what's to come, perhaps an incurable pathology that drives ecocide, geocide, planetcide, a mysterious disease not yet described in the DSM. "An uncanny something seems to block the corrective will, not simply private cupidity or political inertia," Paul Shepard writes (3). What to call it? How about *the human syndrome*? The human syndrome seems to manifest in the following:

> Growth fetishism
> Reality refusal
> Arrogant ignorance
> Nature hatred
> Transcendence dreams

You wonder what could blunt the spread of even one of these interlocking manias. There was a time, perhaps forty years ago during the first flowering of the green movement, when you might have believed that reason and self-interest were enough to set the world on a course away from the self-inflicted asteroid.

But for our sweeping and seemingly ineradicable denialism, only fools and liars would be able to hold such a belief today. Instead, every one of us has become a shill for the human medicine show and its miracle tonic of inane Pollyannaism.

Recently, I dug out an old copy of a postapocalyptic science fiction novel, *Songs from the Stars* by Norman Spinrad, that I'd read as a teenager. The existential problematic it dramatized has always gnawed at me, even thirty-five years on. In the denouement of the novel, Spinrad reveals that iconoclastic scientist Arnold Harker, who with his taboo, "pre-Smash" technology seeks to return inward-looking humankind to space, has discovered an expanding black hole that will devour our galaxy within three million years. This foreknowledge unnerves him to the point where living becomes impossible. "'So that's why Harker killed himself,' [Sue] muttered. 'In the long run, our lives, our dreams, and everything we do for the benefit of our species. . . . It all really *is* pointless, isn't it?'" (26). But the protagonist, Clear Blue Lou, rightly wonders if Harker hasn't taken this revelation a bit too hard, given that he and probably anything that still resembled humanity would not be around to see the catastrophe anyway: "You and me and the people of Earth and all those civilizations out there and even poor Arnold, if he had only understood—we're all in the same place we always were, love. Tell me how this makes anyone's personal fate any different."

Well, you'd probably agree with Lou that Arnold's glimpse of incomprehensibly distant doom changes nothing in the graspable present. It's one thing to know the world is going to end tomorrow, another to know we've got a hundred thousand generations' grace. Most of us can easily accept that humanity isn't forever, and if our demise is projected sufficiently far in the future, why give it any thought at all? For now, the sun will come up and the world will abide. For now, there is today and the joy to be wrung from it.

Yet for those who experience the crushing loss of a child or a sister or a close friend or even a beloved pet, a certain sort of galaxy seems to have winked out: right there and then, not tomorrow or the day after or a million days hence. In time, every one of us comes to know what it *feels* like as such brilliancies end. And right now, right this moment, countless worlds *are* ending under rising waters and heavy weather, with failed harvests and empty catches, and in bloody struggles over what remains. So why, with its vast back-catalog of individual and societal extinctions, does the human syndrome remain unaffected by the continual reminders of the precariousness of life? Why can't civilization conceive of its own death even as death presses in all around? Why does it fail to discern the fatal orbit? Why must we deny the falling skies?

Photo by author

The Hopists, or A Sense of an Ending

> Whatever happened? A breach in the very unity of life,
> a biological paradox, an abomination, an absurdity, an
> exaggeration of disastrous nature. Life had overshot its
> target, blowing itself apart. A species had been armed too
> heavily—by spirit made almighty without, but equally
> a menace to its own well-being.
>
> —Peter Wessel Zapffe, "The Last Messiah"

HOPE AND ITS DISCONTENTS

On May 21, 2011, the world did not end. For those who had staked so much on what was to have been a welcome apocalypse, the failure of its materialization produced confusion and disappointment. But, as has been the case many times before, rarely did the true believers reject the prophet—or the belief system—that had called for it. In fact, when the world again refused to end on October 11, 2011—just as it had not on previously predicted dates of May 21, 1988, and September 6, 1994—the prophet of these gloom tidings, Harold Camping, laid the whole debacle at God's feet: "He allowed everything to happen the way it did without correction. He could have stopped everything if He had wanted to" (Duell and Steven). In other words, it was God himself who had tantalized us with the promise of the Last Judgment, and it was God himself, for reasons only He knows, who moved back the calendar.

Faulty predictions of end times are old hat for these sects, for whom the real test of faith is to see how many times they can put up with having their faith tested. For others, who had no time for the end of the world to begin with, there was chuckling and knowing winks, as if to say, "yep, still a sucker born every minute." But for a smaller subset, a rather tasteless group, frankly, this *epic fail* was more bittersweet, tinged with cheerful disappointment, if such an emotion exists. I am talking about the people who said to themselves of this disconfirmed apocalypse, "well, there goes another missed opportunity." For this group, let us call them the *crypto-apocalypticists*, the End of Days, though shocking and terrible, would at least have constituted some decisive action by our maddeningly indifferent universe. Not to say this group buys into the Christian plotline with its irascible creator who at any moment might pull the plug on his creation. No, it is just that they would not mind a verification—even via catastrophic signal—that the world is less banal and purposeless than it appears. A supernal irruption in the otherwise flat course of events: now here would be something to get off the couch for (though the arrival of Zeus or Odin or Cthulhu would serve as well). Frank Kermode notes that "paradigms of apocalypse continue to lie under our ways of making sense of the world" (28). For good reason: the universe of quotidian experience is bereft of exactly that narratival linchpin that apocalyptic scenarios helpfully deliver, namely, as Kermode puts it, "a sense of an ending." Indeed, history looks more than ever like one damn thing after another, a string of non sequiturs blatted out autistically. A day of reckoning would provide welcome punctuation. Surely all of us, not just the fans of the book of Revelation, would be delighted to learn that our cramped and trivial lives were entrained to some broader cosmic pulsion, and that we ourselves were to witness the grand finale? As it stands, however, eschatology has been replaced by the conveyor belt of linear consumerism: the age of Walkman giveth way to the Age of iPod, soon to dissolve into the Age of iHologram or some equally charmless epoch. Where is it all going, besides through the Apple Store, this long, dull road to nowhere? Economic and technological development proceeds apace, but would not Hegel or Marx be appalled to know that History had gestated neither Absolute Spirit nor the worker's utopia but Coca Cola Zero and *Dancing with the Stars*?

What the crypto-apocalypticist reserves for special condemnation is that most audacious of human values, hope. Hope (styled among the Campingistas as *faith*) is behind the ecstatic pigheadedness of those who never give up on God's always-impending judgment. But it is equally behind the crypto-apocalypticist's willingness to drop in a heartbeat sturdy reason and *hope for the worst* every time he gets wind of a Y2K bug or a Mayan prophecy. Of course, his collapsarian ideations trouble him, for they reveal to him that hope and despair are two sides of the same coin. The pessimist says the glass is half empty, the optimist that it is half full. But if the glass is a poisoned chalice, the differences start to blur. Such is the case with our degraded planet and the growth machine that is killing it. Is the one who hopes industrial civilization will grind to a halt really the pessimist? Isn't he rather the wide-eyed dreamer?

The crypto-apocalypticist wonders why hope was left in Pandora's jar when all the other evils escaped, since hope seems to him like a bird of the same feather, every bit as dangerous as the rest. He disagrees profoundly with La Rochefoucauld, who tells us that "Hope, utter charlatan though she be, at least lures us to life's end along a pretty road" (maxim 168). He concurs instead with Nietzsche, who says of hope that "it is in truth the worst of all evils, because it protracts the torment of men" (45). He understands hope to be mostly delusional, a rote product of wishful thinking and self-promotion. More than the other mischiefs, hope is what spirals out of control, giving heart's ease to a human species that should be absolutely terrified by the abyss it is hurtling toward and desperately about the task of putting on the brakes. Hope lets us off the hook; it lets us observe the ice loss in the Arctic, the drought in the Amazon, the tornado in Arkansas, or the flood in Australia, all the while intoning solemnly: "Well, we must hope for the best." At one time, hope was reserved for probable outcomes, ends that could be brought into existence if honest means were applied and timely measures taken. "Hope for the best" was mere preamble for the more substantive imperative, "but prepare for the worst." But hope has devolved into a puffed-up stance of anticipatory entitlement, as if good things will just naturally come to those with an optative frame of mind—"desire and expectation rolled into one," as Ambrose Bierce puts it. Hope is a form of secular prayer, and, like prayer, it off-loads responsibility onto an external agency—namely, futurity, which is presumed to hold the solutions to problems that would cost the here and now too dearly to face. Hope in this age is a leisure-time emotion, hopelessly bourgeois, hopelessly domesticated by Oprah, Bill Clinton, and Barack Obama, all children of hope, to be sure, but now prophets of hope's unconscionable present, wherein hopes are silver dollars scattered by limousine liberals on the luckless lower orders. (At least with the conservatives you find honest dedication to old-fashioned class war and unrepentant crushing of proletarian hopes. Conservatives can barely be bothered to varnish a central truth of our time: that hope comes to nothing without the money and influence to back it.) Yes, the execrable piety that is hope really is the ultimate guilt-free resource: it costs us nothing, it makes us feel swell, and it sells soap, bibles, and self-help books. It is one of the last and holiest of the sacred cows, and though we are perennially bidden, as Jesse Jackson admonished us in his presidential campaign of 1988, to "keep hope alive!," what passes for hope today is an abomination and should be put down. It is not hope you need when you round the bend in the trail and come upon the bear, or when your tires hit the ice and the car begins fishtailing. Hope is not what you need when the crisis moment arrives and only quick and deliberate action will save you. Hoping for the best can get you killed. At best, hope is an indulgence you might allow yourself when you have exhausted all other means, perhaps as you wait on the roof for the helicopter to lift you from the rising waters. Hope ought to be reserved for the time of no hope.

Many of us have seen Winslow Homer's striking painting *The Gulf Stream*. It depicts a shirtless and barefoot black man lying on the deck of a rudderless, broken-masted sloop adrift on a dark, heaving sea. Sharks roil in the foreground, a water spout leans in from the near background, and far away, hazy and hull-down, a square-rigger in full sail scuds by, unnoticed and un-noticing. With death closing in, the doomed man gazes stoically into the only tolerable direction, which is away from the dreadful scene itself into the mystery beyond the frame. This painting

The Gulf Stream by Winslow Homer. Courtesy of the Metropolitan Museum of Art. Catharine Lorillard Wolfe Collection, Wolfe Fund, 1906.

is usually understood as a portrait in courageous resignation to unalterable fate, a tribute to the peace of mind that supposedly arrives after despair and before termination. But to me, it represents one of those rare moments when we are justified in placing our hopes in unseen things: When all choice is removed, all action foreclosed, then by all means let hope fill the vacuum. Until then, let us be about the tasks at hand.

GENRES OF DOOM

This topos, *hope*, functions in a bewildering variety of genres, from political speech to business proposal to personal narrative to greeting card. And, perhaps surprisingly, it has been a crucial ingredient in the contemporary literature and cinema of disaster, social decline, collapse, and extinction. We will discuss the formal properties of this touchstone in a moment. First let me ask: Does the apocalyptic genre appeal to you? It has long appealed to me—from gloomy youth to gloomy adult. I am not alone: the catalog of popularly successful works in the field of imagined catastrophe dates back at least to Mary Shelley's *The Last Man*, through H. G. Wells, the postwar giant monster movies from the Hollywood and Tokyo film studios, the cozy catastrophes of John Wyndham and CBS's *Jericho*, the grand doomsday guignols of George Romero and Stephen King, the disaster schlock of Roland Emmerich and Michael Crichton, the Christian science fiction of Walter M. Miller and Tim LaHaye, right up to the peak-oil relocalization romances of James Howard Kunstler and George Miller. Many of the top-grossing films of the last thirty years have dealt with destruction and final things: the *Terminator* series, *Independence Day*, *Armageddon*, *2012*, *Deep Impact*, *The Day after Tomorrow*, *War of the Worlds*, *World War Z*.

Is there something about the end of Earth, man, and his many works that speaks to our inner vandal, that part of us drawn to car wrecks and building demolitions, that feels a burst of pride when our country rains bombs on our enemies, that causes children to step on anthills and other kids' sandcastles? In these genres, attraction and repulsion come with the territory, and my sense is that the correct blending of the two makes for a powerful catharsis. Of course, this appetite for imagined destruction is by no means universal. I know plenty of upright people who will not countenance movies and books that depict breakdown and anarchy, and I have taught students who tuned out such materials when I included them in my courses. They do not wish to dwell on these dark matters, and I know how they feel (I'm kind of sick of them myself, after a lifetime of gorging on them). But I also know upright people who never miss the latest apocalyptic film, who are morbidly fascinated by the end of the world as we know it, whether by zombie infestation, solar flare, or some other riveting calamity. Take a look at the teen fiction section in your local public library. You will find that every book series that does not involve adolescent vampirism is apocalypse porn.

This perverse interest is not merely a visceral phenomenon; it wells up from deeper sources. The ongoing production of apocalyptic film and fiction "points to a larger modern syndrome, whose main characteristic is perhaps the constant, obsessive recurrence of the idea of the end," as Matei Calinescu puts it (172). "The end itself," he writes,

> may be regarded as the coming of the *eskhaton* (doomsday in a secular world without transcendence), or the achievement of the *telos* of history (its "final cause," its goal), or a combination of the two in various degrees and proportions; it may evoke anxiety or a paradoxical kind of joy; it may be colored by the most somber pessimism or, again

paradoxically, it may be considered with a mixture of irony, self-irony, and even playfulness. Whatever the case, the inescapable idea of the end remains fundamentally the same, unchanged by the great diversity of reactions, rationalizations, or emotions that it brings about.

Modernity leads ineluctably, Calinescu argues, to a self-critical examination of the foundations of human society; Modernity forces us to cast a cold eye on the very humanism that launched it, the anthropocentrism that maintains it, and the internal contradictions that may bring it down. From Thomas Hobbes to Joseph Schumpeter, from Marx to Margaret Thatcher, the consideration of collapse and disintegration is inevitably bound up in any theory of political economy. This is not to take up the point, with Karl Löwith, that Judeo-Christian eschatology has been secularized into the progressive rationale of the modern state and its purposes, but more simply that the conditions for ideal civic and market operations cannot be posited without the countervailing vision of the conditions for their dissolution. In other words, for civilization to proceed and to smooth out its lumps, there must always be awareness that it could go wrong— and spectacularly wrong. Freud attributed this fixation to the death drive, the auto-destructive force that must be repressed by civilization else its broad-spectrum Oedipal guilt would bring progress to a halt and return us all to some preferential, nonvital state, that is, *dead*. Perhaps the movies and books that express this collective death drive are part of a secular grimoire of spells and incantations meant to cheer our passage through the seductive dark, the societal version of whistling past the graveyard.

At the risk of losing some important granularity, let us say that such films and books fall into two prime subgenres: those that portray the end as a prelude and those that portray it as a coda. We may speak, in other words, of the hopeful doomsday and the hopeless one. In the first category, we have almost all members of the broader genre; in them, the message of hope supersedes the collapse itself. These works stay true to "apocalypse," which means revelation, or, literally, a "lifting of the veil." With the end comes, typically, the truth of things, and with the truth, the chance for a new beginning, perhaps on a more righteous footing. This pattern is easily evidenced by a look at any number of exemplars of the genre: *The Postman*, *Waterworld*, *I Am Legend*, *The Book of Eli*, *28 Days Later*, *Le temps du loup*, *The Happening*, *Wall-E*, *Mad Max: Fury Road* (one could even include *An Inconvenient Truth*). No matter how dashed are the hopes of mankind, all of these films conclude with the promise of a new day: there is a serum, a savior, a techno-fix; a small band of survivors still in possession of the human spirit, the Holy Bible, the healing power of love; a recurrence of reason, a balancing of means and ends; and so forth. The worldsnake sheds its skin and again basks in the sun. In the hopeless subgenre, there are very few takers: *The Last Man on Earth*, *Dr. Strangelove*, *Threads*, *Escape to LA*, *Invasion of the Body Snatchers*, *Planet of the Apes*. I well remember seeing the *Planet of the Apes* when it was first released; indeed, it is the earliest movie memory I have (and no doubt a strange one for a boy of five or six, who should have been immersed in ducks, Dalmatians, and that darn cat, not violent gorillas and sadistic chimps). Charlton Heston on the beach, looking up at the nuked Statue of Liberty, crying out in impotent fury, "You blew it up! Goddamn you! Goddamn you to hell!" Nothing left but speechless humans and malevolent apes. Admittedly, the rest of the series went on to recover the hopes shredded in the course of this all-downhill adaptation of Pierre Boulle's *Monkey Planet* (which holds up, by the way, despite the immobilizing ape masks on most of the actors). Films that end as inhospitably as this one are seldom produced anymore (witness the more or less faithful 2001 remake by Tim Burton, which nevertheless comes off as lighthearted compared to the 1968 version). Contemporary audiences turn away from unrelievedly bleak views of the human prospect.

Two books and their filmic treatments that seem on first blush to exemplify the hopeless subgenre are *The Road* and *Children of Men*. Quick synopses: In *The Road*, the planet is dead, we are alive. A postapocalyptic picaresque, the narrative tracks a man and his son as they travel out of the piedmont and down through an ashy wasteland, where nothing—not plant, animal, maybe not even bacteria—has been spared. They cross paths with a few starving survivors and roving bands of cannibals. The man dies after delivering the son as far as the ocean, where he is rescued (if that is a word that makes sense in the context of *The Road*) by a relatively healthy-looking nuclear family. Film and book end on that note. As a self-appointed authority on this genre, I am comfortable claiming that there has never been a more harrowing treatment of the human endgame than Cormac McCarthy's.

Children of Men: the planet is alive, we are dead. The human race has become completely, inexplicably infertile (shades of "The Screwfly Solution" by James Tiptree Jr., a.k.a. Alice Sheldon). As Lee Edelman puts it, *Children of Men* "gives voice to the ideological truism that governs our investment in the Child as the obligatory token of futurity" (12). In the narrative's present, the last child was born eighteen years earlier, so that the human species is scheduled to disappear when the last old codger succumbs about seventy years on. Absent the prospect of human continuance, the social milieu of *Children of Men* has devolved into a wholly convincing mix of repose, nihilism, and exuberance. I will not recount the plot, but suffice it to say that after much intrigue across a grim and violent Britain, the narrative rounds off with a pregnant mother giving birth to a healthy child.

Now, we know that the rescued boy in *The Road* must still scavenge for canned food and avoid cannibals, and we know that the miracle infant cannot repopulate the earth of *Children of Men*. Their futures, and the futures of their fictional worlds, are not resolved by these final plot points. What transports these two works from the hopeless to the hopeful subgenre is less in the details of the diegesis but in the far more important—if more difficult to define—component of *theme*. Neither *Children of Men* nor *The Road* will ever be confused with "feel-good" novels and films, but there is little doubt that they manage to affirm very powerfully the values of life, love, and hope.

In their adaptations of the source novels for film, directors Alfonso Cuarón (*Children of Men*) and John Hillcoat (*The Road*) made artistic choices throughout that worked to extract and clarify these fundamentally hopeful messages that the novels, despite their rampant declensionist energies, do indeed support. In other words, the kernel narratives of the P. D. James and Cormac McCarthy novels are developed in the films to emphasize not the fragility of hope but its durability. It is worth quoting these directors in full:

> Alfonso Cuarón: I hope young people will see this film. I mean my generation, we blew it. I think we grew up in a world that was pre-idyllic, and we saw the world collapse in front of us and we tried to believe that it was not our fault, that it was not our responsibility. We felt powerless about the situations as if they were very overwhelming and there's a certain sense of guilt involved in the whole thing. Younger generations, they were born in a world that went to shit already so they have a completely different perspec-

tive of what's going on. I really believe in the evolution of human understanding that's happening in [the younger] generation and the generation to come. My intention was to take [the viewer on] a road trip through the state of things and then once you go through this journey for you to try to come up with your own conclusions about the possibility of hope in a world like this. At the end I cannot dictate a sense of hope for anybody because a sense of hope is something that's very internal. We wanted the end to be a glimpse of a possibility of hope, for the audience to invest their own sense of hope into that ending. So if you're a hopeful person you'll see a lot of hope, and if you're a bleak person you'll see a complete hopelessness at the end. (Quoted in Guerassio)

John Hillcoat: But the one thing that I love about McCarthy's writing is how he creates extreme worlds where people are under real pressure. And that world is as much of a character as the main characters. Also it's an unflinching look. It's a bit literal to just say it's bleak and dark. Hope is all the more special when it's surrounded by hopelessness. . . . I think there is an essence in humans. Basically, when you strip everything bare, when you take humanity down to zero, put it under pressure, like a scientific experiment, I think you will find that it's that kind of hope and those higher elements of humanity that actually saves the man and pulls him back into check, and also gives the boy enough of what he needs to carry on. I think that's part of the genius of the tale. There's an emotional truth to it. We've all experienced, in our own ways, the best and worst in people and even in ourselves. And it's that human strength that's the key. (Quoted in Di Giacomo)

One needs no special training in discourse analysis to understand that for both directors a key goal was to reserve in these dark works a space for hope, and their articulation of this goal is tantamount to them declaring that the films are hopeful works indeed. Yet I do not think that keeping those hope-spaces available was particularly difficult. Today, any author or filmmaker must actively, scrupulously work against hope to keep it out of their projects because hope worms its way in through the narrowest cracks. And this challenge of *hope proofing* is rarely taken on; it is risky and unrewarding and probably tests poorly with audiences. One might say that foreclosing the option of the hopeful ending is the third rail of mainstream filmmaking and publishing. *Children of Men* and *The Road* come intriguingly close to this rail but decline to grasp it.

THE CULTURAL LOGIC OF HOPE

Regardless of their roots, negative utopias or blasted futures like those portrayed in *Children of Men* and *The Road* are now part of our cultural programming, and the constant rehearsal of doomsday narratives has given us a certain taste for them and some predictive insight as to how they ought to unfold. More crucially, the end of the world has become a powerful fantasy theme in late capitalism, which exploits apocalyptic thinking to buttress its own creative/destructive dynamic and to reaffirm the phoenix forces that we *must* believe in to, in turn, believe that our current malfunctioning world system can actually carry on. To put a finer point on this, let us note that at every moment our consumerist economy goes about its task of shaping our desires, supplying them, and then—even as that which was most recently desired is still doing its scut work of reproducing the conditions of its reproduction—casting about for new desires while shunting the old ones toward the smoldering rubbish tip. I believe the end of the world scenarios of popular culture serve to extend this sequence of desire toward the limit situation: What would happen if *all* that we loved went onto the pyre? Every last bit of it. What if the whole system of desire itself was trashed? Where would we turn? How would we carry on? And, importantly, how could we get back what we had lost? End of the world fantasies tutor us in the impressive scope of our desires and discipline us to fear their diminution.

There are at least two ways the limit situation can function to preserve the status quo by ideologically recontaining these unsavory simulations. The first, superficial, superego-enhancing function appeals to what some believe is our innate bright-sidedness (see, for example, Tali Sharot's *The Optimism Bias*): End of the world scenarios remind us that we have never had it so good. In effect, we are put into the position of affirming that, granted, while ours may not be the best of all possible worlds, it is the only one we have and therefore precious and worth defending. What my dad (and probably yours) used to say applies: "Cheer up, things could be worse." It is status quo politics at its most statically quotidian. The preservation of the world as we know it was, perhaps disappointingly, the moral Cormac McCarthy himself attributed to *The Road* when he spoke about it in a rare television interview: "Simply care about things and people and be more appreciative. Life is pretty damn good, even when it looks bad. We should appreciate it more; we should be grateful. I don't know who to be grateful to but you should be thankful for what you have" (*Oprah*). I hesitate to let McCarthy's advice stand for the superficial function, in part because I think his was a shallow answer in a shallow context. But, at the same time, that shallowness is precisely the point: We know exactly what McCarthy means because this sort of quasi-Panglossian rhetoric already informs so much of our daily social and psychic programming.

The second function is more interesting because it operates in the deeper recesses of consciousness yet manifests itself, strangely, as a Rubicon beyond which our thoughts quite consciously will not go. This function involves the acknowledgment of our global civilization's dangerous trajectory combined with its simultaneous disavowal, an unwieldy synthesis made possible through the intervention of the concept of hope, which, as we have seen, is the power-bringing magic word that helps make the bad stuff go away.

Let us approach this function concretely. The basic structure of popular doomsaying can be easily tracked across a number of sites. For example, many of the most powerful progressive voices on the climate change front claim we can save the climate-as-we-know-it by tweaking the current economic model to promote and reward so-called green production and consumption. This more-of-the-same-only-smarter approach was articulated forcefully in the columns of Paul Krugman, the PowerPoints of Al Gore, the rallies of Bill McKibben, and incentivized in the real-world policies and programs of the Obama, Cameron, and Merkel administrations, where selective investments and subsidies promoted windmills, electric cars, solar technologies, improved energy infrastructure, clean coal research, and the like. "Greed is going Green!" What could never be acknowledged, however, is that talk of *dematerialization* or *decoupling* is a red herring, for only colossal, absolute reductions in production and consumption have any chance of even slowing the rate of acceleration of climate change, let alone halting it. The climate system is already loaded with so much inertia that we could reduce the human carbon footprint to zero tomorrow and we would still undergo catastrophic warming over the next century. All of this is impossible to confront, both politically and psychologically. So the basic structure of desire is as follows: We cannot live in a destroyed biosphere; economic expansion is destroying the biosphere; therefore, to save the biosphere we must expand the economy. Of course this formula makes no sense. But rather than follow the logic through its full implications—implications that go to the heart of our failing tenure on this planet—we ignore the premises and revert to magical thinking. This faulty syllogism reminds us of the paradoxical injunction of the schizophrenic narrator in Beckett's *The Unnamable*: "You must go on, I can't go on, I will go on" (179). If, as Fitzgerald writes, "The test of a first-rate intelligence is the ability to hold two opposed ideas in the mind at the same time, and still retain the ability to function" (69), then we must conclude that industrial civilization is pure genius indeed. But something here does not scan. Fitzgerald continues: "One should, for example, be able to see that things are hopeless and yet be determined to make them otherwise." Thinking of his own troubled times, Pablo Casals, the great Catalan cellist, is reputed to have echoed that formulation: "The situation is hopeless; we must take the next step." But here is where industrial civilization is only second rate and, properly speaking, insane: it refuses to take the next step, refuses to do otherwise, instead performing its bloody work over and over again the same way, each time expecting a different result, to paraphrase Einstein. It seems that a two-headed rhetoric is in play, one that is pleased to call regress, progress; catastrophe, triumph; excrescence, growth.

Thomas Homer-Dixon, in his mostly grim *The Upside of Down*, dubs this kind of Janus-faced procedure *catagenesis*, whereby even the wholesale collapse of civilization can be rebranded as an opportunity for new growth and life, akin to wildfire in an overage forest but with none of the latter's ecological legitimacy. As Homer-Dixon describes his catagenetic view, we notice he follows a well-trodden rhetorical path:

> The alternative approach [to standard environmental management] I advocate requires us to adopt what I've termed a prospective mind. We need to be comfortable with constant change, radical surprise, even breakdown, because these are inevitable features of our world, and we must constantly anticipate a wide variety of futures. With a prospective mind we'll be better able to turn surprise and breakdown, when they happen to our advantage. (268)

His argument, though courageously grounded in the hard truths even now too few of us are prepared to contemplate, nevertheless concludes with the same old bromide of politicians and business leaders alike: tell the unemployed, the downsized, the dispossessed, the futureless that when the world hands you lemons, make lemonade. Variants of this can-do rhetoric were deployed by Bill Clinton in 1992—epitomized in his first inaugural address as "the urgent question of our time . . . whether we can make change our friend and not our enemy"—and equally by George W. Bush in 2000—embodied in his own first inaugural as a quotation attributed to Jefferson: "Do you not think an angel rides in the whirlwind and directs this storm?" It is a winning commonplace because its inverse—change is frequently our enemy, the universe is indifferent to our struggles, and things can indeed get much, much worse—is a message our bullish-on-tomorrow culture is spectacularly ill-prepared to receive.

INOPINATUM

How much like the paradox of hope that we have already discussed! You cannot live without hope, and when all hope is gone, hope is what you must cling to ever more tightly. In *The Road* and *Children of Men*, this structure of hope seems very much in play, despite the methodical destruction of hope throughout both films. Here, the saving injunction to "keep hope alive" (recall that the boy finds a new family and the mother brings new life into the dying world) by holding fast to the promise of a better tomorrow is the sop that parallels, nay instantiates, the capitalist program of turning any misadventure or even unmitigated disaster into an opportunity for growth and profit: run a tanker aground but create jobs in the cleanup sector; catch all the fish in the sea but increase sales for salmon farmers; roast the planet but make a killing in air-conditioning. Earth remains profitable even as the economy externalizes life itself.

In dystopian works like these, the schematization of hope seems to correspond to what Quintilian called *inopinatum*, a form of paradox in which the rhetor articulates the unthinkableness of a particular case so as to disarm it (Sonnino 113). Johannes Susenbrotus described it this way: "We put forward a suspicion or opinion by denying that we could have been capable of it. . . . We use words that express wonder at this state of affairs" (qtd. in Sonnino). In the last century and this one, we've seen this rhetorical form used a lot, as in, say, "I cannot believe that the human race will destroy itself." In this example, *inopinatum* heightens the sense of indignation that such an issue should even be raised, and that the values of human wisdom and imperishability should ever fall into such disrepute. It may also have the character of *sustentatio*, or suspense, in which the question's proper answer—that is, "no, of course this cannot be the end of the human race"—is held back until the completion of a (lengthy) preamble during which the audience is forced to walk with the rhetor down that unthinkable road.

In fact, the collapse of industrial civilization is quite thinkable. It lies before us as a distinct possibility. The premises of these novels and their filmic versions are very much inspired by today's eco-catastrophic headlines; they could easily be excused were they to conclude on the dark and discouraging notes that so many of the reports coming in from the field do. Nevertheless, in these austere declensionist novels and films, the figure of hope, foiled at every turn in the body of the narrative, is reasserted powerfully at the end—almost as if to say that despite appearances the work was never about hopelessness but about hope's eternal return on the shoulders of an indomitable humanity. In this ideologically saturated age, ensconced as we are in hope's high temple, that is the message we long to hear and, to our great peril, the only one we seem willing to hear.

Inopinatum: the declaration of a truth that one recants in the same breath. If you think these books and films are examples of this Pollyannaish core in even our most unflinching prognostications, then you, like me, must be aghast at what we are telling ourselves. For let us not pretend these cultural works, created to feed the insatiable appetite for doom and gloom, are messages from a future that must be, or even presentiments of a probable one. No, they are made from the values and aspirations of our time, and they trace out our own axis of despair and hope. What is most terrifying about them is that they reveal our concept of hope is a fraud, a shiny apple hiding a worm at its core. The appeal to hope is simply the growth machine voicing through us its impossible, reckless vision of limitless expansion, another ruse by which our suicidal culture asserts its authority over our imaginations and robs us of our ability to come to grips with its perverse trajectory.

Photo by author

The Decivilizationists, or Impure in a World Unpurged

> Man, his loose moods disjoin; madness is under the skin.
> —Robinson Jeffers, "To the Story-Tellers"

> We are born mad, acquire morality and become stupid and unhappy. Then we die.
> —M. D. Eder, "The Myth of Progress"

THE BIG C

Why me? What did I do? How do I fix this mess? You mull over the Bigs. We all do, most often in those low moments when the velocity of living is slowed by grit in the workings: say, a crappy day at the office. What's it all about, Alfie? Wouldn't it be nice to start effing those ineffables, especially when you're down in the dumps and could really use a lift? From the bottom of a well the sky looks blue and inviting, despite being farther away and a helluva lot smaller.

One thing's for sure: even in the darkest depths it's easy to see we're trapped in a world-girdling meat grinder that's turning everything into sausage. (In this metaphor, by the way, you're both the meat grinder and the sausage.) It takes no special vision to glimpse our general horribleness, not in this eleventh hour when our teensiest impacts on the earth threaten a tipping point. Say what you will about zero emissions and carbon-free lifestyles, each one of us is a matter-energy sink, a minivortex that wobbles around the yard vacuuming up plants, animals, soil, air, gasoline, and eventually, the wife, the kids, and the family pet. We suck, basically.

At the same time, we blow. Out of us spews—from the usual places, of course, but also from other exits hidden from view—the wastewater and heat, the toxic stews and brews, the stripped and sawed-off molecules. All that effluvia lies or flows or drifts around us. Tough luck we can't reuse it, but we want the virgin, negentropic stuff. We're not mushrooms, for chrissake. Anyway, I wouldn't want to be the custodian of this planet after we've finally killed the lights and made our way off the blood-drenched stage.

Still, you get to thinking about our similarity to fungi, building up their spongy order on rotting logs. And we can be compared fruitfully to many other organisms: sharks, wolves, vultures, viruses. But such metaphors are unflattering to those beings. Better to compare us only to ourselves. We are who we are and that's more than enough.

What, exactly? We are mostly that entity too big for its britches. We run XXL in all directions. The planet won't fit us anymore, and its seams have already been let out all the way. We've grown fat off the land, and now there's not much left but the marrowbones. What we fondly call civilization has been the Haystacks Calhoun of complex systems, a clumsy but unstoppable bulk that pile-drives everything and then smothers the life out of it. We've become the super-heavyweights of creation.

Indeed, there are some who call for an end to civilization; they claim decivilization is the only route to getting ourselves right with the planet. It's civilization that's been causing all the trouble: dams, factories, cars, coal mines, pills, pop cans, plastic bags, trips to the mall, trips to the dump—all of those are manifestations of the core disorder that is the Big C itself. Folks with this view believe the answer to the problem of civilization is to bring the whole edifice down, like a slickly demolished building that collapses without taking the rest of the block along with it. With civilization reduced to rubble, the rest of creation could heal. So say the decivilizationists.

TAKING IT DOWN

Perhaps it troubles you to read such blasphemies. Civilization, the cause of all the world's woes? That claim is rather sweeping, not a whole lot less grand than blaming death on life. Using this kind of generalization, we would be hard-pressed to disagree with the following: "Do you know what I attribute the world's disorder to the breakdown of? Pangaea! If only the continents had stuck together we wouldn't have had so many differences in the first place." And, on the other hand, isn't civilization all about managing such differences, through, as Fernández-Armesto puts it, "the civilizing ingredient" (30)? Civilization achieves the integration of peoples via its homogenizing appurtenances: religion, commerce, art, science, technology, the city, the state. Civilization provides common frames of reference and shapes our shared aspirations for knowledge, transcendence, comfort, fashion, pistachio ice cream, Kanye West music. Civilization creates the conditions for assured survival. If civilization were to come down—if it just fell of its own accord or if someone from the Earth Liberation Front or Al Qaeda gave it a push—what would we be left with? A hardscrabble existence in the twisted wreckage. Untreated sewage. Stray fissionable materials. Cholera. Starvation. Mad Max. War of all against all.

Anti-decivilizationists will argue, "The fact that civilization produces great ills cannot be an argument for civilization's abolition. We must *complete* civilization. We must civilize our way through this extraordinary period of adjustment." Or they might explain, "Look, it's not civilization per se that's to blame. It's the current *mode* of civilization that's the problem. The Incas were fine, as were the Mycenaeans and, umm, the Hawaiians.

The Edo period in Japan was darn-near steady-state. And let's not forget the Hopi. Those were sustainable civilizations, living in harmony with their environments. Sort of. Maybe. We'd like to think so. In any case, civilization can be made to work. We just need to shift our modern civilization onto a firm ecological footing. Get inputs and outputs balanced, sources and sinks aligned, mind and nature attuned. Maybe geo-engineering'll have a role. Assuming that's an actual thing. Well, whatever, it won't be easy, but history will judge us. So let's get busy!"

Not so fast, says Derrick Jensen, the most articulate and indefatigable of civilization's gadflies. In a number of volumes detailing the dismemberment by humans of the planet, Jensen has argued for the absolute necessity of bringing to a close the reign of modern industrial civilization. If you want salmon, whales, and forests, civilization must go. If you want clean air, clean water, and clean food, civilization must go. If you want a future for yourself, your children, and your grandchildren, civilization must go. The octopus arms of our great industrial civilization are squeezing the life out of all they touch. And there's not any single failure to point a finger at; you can't smooth off a rough edge here or there and expect civilization to shape up. Instead, civilization produces a full-spectrum malignancy that does grievous harm to all other things—including the social webs that bind humans in love and harmony with each other. For Jensen, civilization is irredeemable; it is rotten to its core. We have a moral imperative to pull its plug:

> If civilization lasts another one of two hundred years, will the people then say of us, "Why did they not take it down?" Will they be as furious with us as I am with those who came before and stood by? I could very well hear those

people who come after saying, "If they had taken it down, we would still have earthworms to feed the soil. We could have redwoods, and we would have oaks in California. We would still have frogs. We would still have other amphibians. I am starving because there are no salmon in the river, and you allowed the salmon to be killed so rich people could have cheap electricity for aluminum smelters. God damn you. God damn you all." (*Endgame* 1: 93)

So civilization must be defeated if the planet is to be saved. But how do you go about doing that—the *taking down* part? That seems like an operation that'll need more than a few jars of elbow grease. Well, Jensen's view is that it's going to happen anyway. With industrial civilization so manifestly unsustainable, it's only a matter of time before it crumbles under its own weight. But before it does, civilization will have rubbed out most of the planet's species and life support systems. Therefore, Jensen urges us to apply force to civilization's pressure points *now*. Doing so would involve selectively targeting infrastructure like electrical grids, computer networks, roads, and bridges—the circulatory system of the behemoth, if you will. Civilization can't do its bloody work if its arteries are clogged. To get the job done right we would need some hackers. We would need truckloads of TNT. Dams are a particular sore spot for Jensen. Living in the Pacific Northwest, he sees firsthand the damage they do to rivers, the creatures that live in them, and the indigenous peoples who rely on them. He prefers dam removal through legal channels but will accept any means necessary. "If you're like me, you're probably wondering how much explosives it takes to knock out a big dam. The answer may make you as happy as it makes me. It doesn't take that much at all" (*Endgame* 2: 595).

ALT.CIV

For most of us in the cattle car of civilization, the memory of an *outside* has faded to black. Civilization is the whole shebang. What's its alternative? Some awful, unthinkable badlands, hardly conducive to our preferred lifeways. All we know of hunting and gathering we've learned from *Naked and Afraid* and *The Walking Dead*. Not tempting models. We've grown used to our confinement and can no longer imagine a pleasant walk through the tall grass past the buffalo herd down to the watering hole.

But Jensen wants to rekindle our imagination with his vision of a better world, a greener, grander world that brims over with life as it last did long ago when civilization was still in the cradle. His incendiary polemics combine with passionate evocations of lands and seas repopulated by nonhumans; he triple-dog-dares us to recover that better world from the tainted and barren sacrifice-scapes produced by our suicidal industrial regime. His is a vision of small, local economies rooted in subsistence farming and natural bounty. People would live their lives enfolded in nature and community, not wrapped up in fantasies of material gain and techno-dominance. A New Arcadia.

Jensen's many books, like Edward Abbey's for a previous generation, appeal to the armchair radical. Literate, loquacious, and furious, Jensen provides his readers with the thrill of direct action without picking up an actual monkey wrench. You read Jensen and come away simultaneously defeated and energized. In the end, however, most of the concerned citizens who digest his volumes will conclude that his cure is as awful as the disease. Emotionally, one can be swept along by Jensen's argument, but intellectually the claims are impossible. Equally impossible are satisfying answers to the questions a reader must inevitably ask of him:

What will happen to the bodies? The violent removal of the pillars of industrial civilization will instantly create astonishing shortages of every necessity, but above all, food. Agriculture is the original human sin, if you will, but its suspension will not lead to our redemption. When the oil-based ag-complex is shut down, old-style muscle-powered cultivation will not do the trick. Rather, famine will immediately touch all but the elites (who doubtlessly will find ways to secure food supplies in the immediate aftermath). The disruption of the food distribution network will sentence most urbanites (50 percent of the planet's population) to starvation. None can feed themselves and most have only a few days' supply of food in their larders. The endgame of civilization thus involves a massive die-off, which Jensen himself acknowledges. The stench and corruption from the dead will overwhelm cities, likely rendering them uninhabitable. Disease will spread, affecting those not already starving. Survivors will surge into the countryside, picking clean the land like swarms of locusts. (As only one of many disturbing side notes: What's to prevent zoo animals from being devoured, and every remaining salmon yanked out of its stream and eaten raw? Hollywood has made this movie dozens of times before.)

What to do with the machines? The nuclear reactors, the refineries, the pipelines, the coal-fired plants, the supertankers laden with oil, the sewage facilities laden with sewage, the mining and industrial waste ponds already only flimsily retained by cheap and porous barriers? Who are the skilled technicians who will take all that infrastructure down safely as they are scrambling for their lives? Who will tend to the tons of radioactive materials we've produced and stockpiled? (About nuclear plants, Jensen asks an engineer, who finally admits, "I don't have a good answer what to do about it. When you let the nuclear genie out of the bottle . . . you get yourself into some deep shit" [*Endgame* 2: 843]).

What about the places (Afghanistan, Somalia, Syria) where civilization is already in the state Jensen prefers: anarchic, minimalist, local? They provide small previews of a decivilized world. What makes him think postapocalypse America will go gently into the long night, safely decentralizing into small ecotopian jurisdictions, like those presumably waiting to form in Northern California? Will Alabama and Arkansas suddenly burst forth with intentional communities? Will America evolve toward postindustrial pastoral simplicity—or devolve into checkpoints and chieftains? As the preppers say, when the S.H.T.F., and we are left without rule of law (WORL), it's TEOTWAWKI.

And what about women? Women, whose aspirations are already denigrated and denied in this dangerous man-cave we call civilized America? If there is anything Jensen has been more impassioned and eloquent about than the plight of the earth, it's been the abuse of children and women by their fathers and spouses. He knows of it firsthand. But without police and laws, support organizations and standards of decency, can we expect all men to become paragons? Isn't it at least as likely that for women, a central concern in postcivilization America will be avoiding rape gangs? Slave gangs? Jensen will point to the observation that hunter-gatherers are more egalitarian, and rape among them is virtually unknown. Let's stipulate to that. But in the meantime, can we really believe there won't be many horrific speed bumps on the road to the New Arcadia?

And what of those who came late or unwilling to these shores? Those with no established birthright, who maybe never embraced the nation wholeheartedly because for them it was too often a nightmare? Who therefore might be seen to have caused this painful fall? Those whose complexions are not pale, lineages not pure, thoughts not straight? What makes Jensen think the usual scapegoats will not be ratted on, rounded up, and put down? What makes him think that? Surely not an attentive reading of human history.

And what of the components of civilization worth preserving? Music, painting, films? Poems and plays? The elegant line and the dramatic pause? So long ballet and opera. Nice to have known you Boston Pops, MOMA, and Smithsonian. If you don't care for literacy, that's great, because you won't need it in the new world disorder. Your Sunday morning newspaper has suspended delivery. Permanently. Are you a fan of health and hygiene? Do you value the tools and expertise civilization has developed to combat infant mortality, treat mental illness, and provide geriatric care? What will you do when you need a root canal? A flu shot? A vial of insulin? Are you prepared to be crippled by a torn ligament? To die from an infected hangnail? To murder for a tube of ointment? Jensen will say such perils are locked in anyway as the biosphere unravels; bite the bullet now and perhaps we can cherry-pick a few essentials to take with us into *alt.civ*. Better to follow the path of conscious collapse than stumble blindly into the catastrophe.

All in all, Jensen's vision of a soft landing after the initial shock belies the track record of large-scale societal breakdown. The aftermath of World War II in Europe involved massive internal migrations, widespread violence, disease, and hunger, even as occupying powers attempted to restore order. When civilization is taken down, with no central authority to reign in the chaos, horror will compete with horror. All bets are off. You may not like the meat-grinding aspect of civilization, but there is a concomitant element, *civility*, that's got to be worth something.

ROUND UP THE USUAL METAPHORS

I'll put my cards on the table: I fully agree with Jensen's basic premise, that civilization, a complex system that has done untold harm to the living world, is on its way out. Infinite growth on a finite planet doesn't wash. The correction is coming, and it's going to be agonizing. Still and all, I must describe my declinist ideations as running both shallower than Jensen's and deeper.

Shallower: I just can't put my seal of approval on the chaos and loss of life caused by the violent, intentional withdrawal of civilization's life-giving appurtenances: hospitals, medicines, schools, grocery stores, civic governments, water and sewer, electricity, telecommunications, news organizations, national parks, game wardens, fire departments, law enforcement . . . well, what I think are essential services comprise a lengthy list. We are asked by Jensen to imagine living happily in a restored world that we obtain by condemning billions of our fellows to destitution, misery, and death. He outlines a Sophie's choice that very few of us could ever make. In a nutshell: it's hard to get behind a playbook that would need to include a chapter titled "How to Tell Your Children Daddy Blew Up the Electric Company."

Deeper: Unlike Jensen, I see no realistic solution to the problem of civilization, not even actively "taking it down." Jensen, as it turns out, is far too optimistic about our ability to topple the edifice we have built up over millennia. The program of civilization will not meekly permit itself to be uninstalled. It has its own defenses. It fights back. It is a superorganism with a robust immune system: a set of seven billion antibodies who will coagulate around its wounds. Civilization will rebuild broken dams, repair fallen electrical wires, and defend against system hacks. It has much redundancy; it can survive the surgical removal of many of its parts. When civilization fails—and I agree that it will fail in all sorts of unanticipated ways, long before it comes close to *completing* itself—it will not fail because it was felled by Lilliputians but because its own weight bore it down.

Let me trot out a well-used metaphor: When we signed up for civilization, long ago in our misty past, we were boarding a gigantic, badly designed ocean liner. Back then, if the ship struck the reef on the way out to sea, we were close enough to shore that maybe we could swim back. But as we traveled beyond the horizon, our options narrowed. The "back to the land" possibility was foreclosed, and abandoning the ship wasn't an attractive option. The vast majority of us depended on keeping this leaky behemoth afloat and ensuring it cruised forward. If we shut down the engines, the pumps would fail and the vessel would sink.

Centuries, even millennia, have passed since our departure. And now we're sailing the vast, open ocean. There's nothing but the angry sky above and five miles of cold, black water below. Civilization can't be taken down because there is nowhere else for its passengers to go. If it goes down, we all go down with it.

And complicating things, of course, is that damned iceberg dead ahead.

I'm a passenger on this clichéd ship, locked down in steerage with the rest of you. (I regret to report that those on the upper decks are laboring under the delusion that they're not subject to the same forces that we are, that they can somehow secede from the biosphere, maybe catch a helicopter to New Zealand and ride things out in their doomsday redoubts.) I agree with Jensen that there is no *technical* solution to the problem of civilization. The only remotely plausible fix has only ever been a *political* one, which at this point means a near-instant commitment to an accelerated degrowth pathway whereby carbon-based energy is zeroed out in one or two decades. In other words, the entire developed world would have to learn to live like the Amish in a very short time. Needless to say, this massive shift to a low-energy world is not going to happen any time soon. We are not going to leave the oil in the ground, we are not going to give up our machines, and we are not going to put the brakes an economic system that reveres, rewards, and requires growth. We're going to keep doing what we're doing until the world we've always known is no more.

PLEASE, MAY WE CHANGE THE SUBJECT?

If dwelling on the sterile prospect of rapid degrowth gives you a headache, perhaps you'd like to sift through a different collection of unhappy thoughts? Answer the following question: How is it that you're able to mentally cope with the ongoing disasters industrial civilization metes out? It's not enough to say that on the great ship we get three squares a day and unlimited shuffleboard, and that all that food 'n' fun diverts us from the pickle we're in. To live with a contradiction is one thing, but to live under a cloud that rains contradiction perpetually is another. We ought to be going through life in a kind of mental fetal ball. Jensen says that in modern society everyone suffers from a form of complex PTSD: "The entire culture is so violent, so traumatic . . . as to render most all of us to one degree or another shell shocked, and therefore incapable of realizing or even imagining what it would be like to live a life not based on fear" (*Endgame* 1: 70). So what's our secret for keeping ourselves from always running around as if our hair is on fire?

There appear to be three main defenses the human brain can mount against the boomeranging trajectory of civilization. Any person could find herself in one of these states of mind depending on how deeply she has been injured by civilization in the course of her day. The first defense is the attitude 99 percent of us adopt 99 percent of the time as we stumble through life: submit to the status quo or at least to some negotiated version of it. This is the "avert your eyes" school of sanity management, little more than a fugue state in which the wounded psyche flees the damning evidence of planetary dysfunction and goes to its happy place of palm trees and piña coladas. Most of us like things the way they are, or have convinced ourselves that we do, and we'll fight tooth and nail to preserve our pleasures and privileges as we currently enjoy them. Even those of us who know we are endangered by these selfsame privileges will adopt the status quo ideology in order to imagine we will be whisked out of harm's way by the status quo reality—which, sadly, serves only to reproduce its failures and lengthen its period of ascendency. Hegemony is a bitch, to paraphrase Lenin.

Structurally, then, as the successive personifications of a militaristic petro-state, Obama and Trump weren't so far apart on these matters, both of them pushing the toothpaste around inside the same tube, which is to say the centuries-long heritage of unsustainable growth and resource overexploitation. That Obama owned up to the hazards and Trump denied them did not change the paradigm appreciably; both were captains of our unseaworthy ship heading for its icy rendezvous (granted, the appalling Mr. Trump wore Thurston Howell's yachting cap and gleefully laid on more steam). Well, they joined the rest of the 99 percent in hewing to what is believed about how civilization works: It will work until the day it doesn't, and we trust that day won't arrive until well after our own personal business here on Earth has been concluded.

A second mental defense is the one adopted by folks like Jensen, Theodore Kaczynski, the doomsday preppers, fans of the book of Revelation, and the other members of this tiny minority who have come to view civilization as a total boondoggle and will be glad of the chance to give it a boot in the backside as it leans over the cliff. As I've already outlined, the decivilizationists would willingly roll the dice on a better tomorrow by burning down today. Now, most of us don't want to take that bet; to lose is to lose everything. And so we remain civilization's faithful lapdogs. By contrast, Jensen et al. figure the abrupt landing at the bottom can't be worse than the deadly shenanigans at the top. (The rest of us see our own twisted corpses at the foot of the bluff and shrink back from the prospect.) So the question here becomes something like this: What goes through the head of the decivilizationist as she maintains her anxious vigil on our slow-motion collapse? Unable to block out the dire evidence of an unraveling planet, what lets her sleep at night?

Decivilizationist protective cognition seems rather perverse: the mind simmers with righteous anger at the unbearable present—and then it is soothed by fantasies of postcollapse regeneration. Anger energizes, but only for so long, and so sooner or later the mind must turn to a virtual resolution. Aristotle claimed our emotions work in pairs, and a key to effective persuasion was to move audiences along the continuum from one pole to the other in a kind of catch-and-release strategy. Obeying the general principle that we seek pleasure over pain, a good rhetor will bring a negative feeling into play, then remove it, jockeying his relieved audience into a state in which the emotional ledger once again balances. In the same way, the decivilizationist, in an act of communal or self-persuasion, loads up on ecological horror in advance of the cathartic postcivilization imaginary.

Needless to say, this pulsating, internal tug-of-war can hardly be a recipe for equipoise over the long haul. Burnout, if not outright catatonia, is a distinct possibility. The third mental defense thus seems more sustainable and perhaps is what exhausted decivilizationists (not to mention the deluded 99 percent) must ultimately take up as their coping strategy. At present, it's the characteristic stance of only a small minority. Articulated by collapsarians like Roy Scranton, Carolyn Baker, and Guy McPherson, it forwards the notion that the earth has already become a very large, open-air hospice. These writers explain that since our civilization is already doomed, that the disease of self-destructive progress is in its final stages, our primary task is to come to terms with our descent into the long goodbye. What they're advocating is a state of dignified acquiescence, something like the attitude of a terminal patient resigned to his condition and at peace with his fate.

Deference to terrible destiny. Ennobling triumph of spirit over corrupt flesh. Yet it's a state of mind most humans will have trouble cultivating. It demands a surety of self and of universe beyond the reach of the average brain, and an equanimity—in the face of monumental stupidity, greed, and villainy—that only saints commonly achieve. What would it look like, feel like, to adopt this amazing grace and grit? Baker explains:

> In these last hours, here are seven things we must be doing: 1) Falling in love with nature and allowing our hearts and bodies to weep with it; 2) Allowing that relationship to make exquisite meaning in our lives and determine everything we do; 3) Preparing with unprecedented awakeness, emotionally, spiritually, and logistically for death; 4) Doing everything humanly possible in our sphere of influence to practice good manners with other species and soften the impact of their demise; 5) Reflecting on our life in terms of how we lived it—what worked and what didn't work, what we wish had been "different, better, more"; 6) Making amends with people we have harmed; 7) Cultivating compassion for those who harmed us; 7) Creating beauty, magic, and joy in our lives as often as possible with as many people as possible. (Baker and McPherson)

Easier said than done. My sense is that if humans were equipped to encompass these seven virtues our species would not be in its current fix. Nevertheless, I expect the hospice stance can only gain adherents as we move deeper into the crisis.

Here is a question some of us would like to pose: Is there a fourth option? A useful mental calibration that is not denial, anger, or acceptance? I'm wondering about curiosity. Perhaps there's a way to reduce the angst of having to watch suicide-by-civilization—to take a keen, naturalistic interest in the demise of the human-trammeled earth—without getting all bent out of shape over it. Because if that stance were possible, wow, that would definitely soothe the faltering psyche. There are nuances of feeling to be explored here. First, curiosity does not imply joie de vivre. On the wall of my brother's fish market is a plastic largemouth bass fitted with a motion sensor. Every time you pass near, it waggles its head and starts singing Bobby McFerrin's "Don't Worry, Be Happy." (So advised, I confess I have thought about swapping anxiety for opioids.) Yet what a contrary, mean-spirited approach, to wake up every day and enjoy the descent into eco-catastrophe! Second, curiosity does not imply fatalism. I have no use for folks who say things like, "Well, in the long run, we'll all be dead," or "In the long run, the planet has seen many catastrophes, extinctions, and upheavals, so don't fret that humankind is now being broken on the wheel of life." Those sentiments are trite and self-serving. Really, how could you look with a jaded eye on dying seas and spreading deserts, on melting ice sheets and environmental refugees? You'd have to be a strange cold fish to remain nonchalant about the auto-da-fé that is being held every second in this, the sunset hour of human ambition. But let's think about this concept of bearing witness, and of collecting, with a certain sangfroid, accurate notes on the end-times: There's surely satisfaction and some comfort in *getting the details right*. I suppose the mindset might be akin to that of the pioneering surgeon who removed his own appendix to prove the efficacy of local anesthetic. It's not as if you're experiencing something pleasant. But you're confirming a theory, following through on a plan, doing your job. You may not be able to slow the downfall of humankind, but at least you can check off the boxes as you spot the signs and portents.

No dice? You're not buying it? You say there's nothing more objectionable than lofty reason looking down its nose at evil in the flesh? Well, perhaps there is yet another stance, one that doesn't call for brain-dead denial, counterproductive violence, impossible composure, or compensatory distancing. How about this: an active withdrawal of services, a kind of mental sit-down strike against the meat grinder? Mulish refusal accompanied by sardonic laugh. I have in mind an image of Galileo muttering to himself, "E pur si muove." What would it mean for us to similarly refuse to collude with the delusion, adopting instead a Bartlebyian attitude of "I prefer not to"? Weapons of the weak, as James Scott dubbed them, are the quotidian tactics of oppressed peoples in societies where open dissent is not possible. By any other name, we're talking about *peasant resistance*: It includes "foot dragging, dissimulation, desertion, false compliance, pilfering, feigned ignorance, slander, arson, sabotage, and so on" (xvi). Now, I'm not interested in the weapons (though foot dragging resonates with me a lot) so much as the attitude: aware of the folly, certain of the crisis, deaf to the lies. Not hopeful, yet not depressed. Not frantic, yet not apathetic. Not continually opposing, yet never rolling over. Remember that I'm trying to frame a state of mind here, a thought-technique for riding out the shit storm. For that we require a resolve adequate to the challenge of having to face the days, the weeks, the years, and the decades of worsening ahead. We might call this mental posture *hunkering down*.

For what? Why hunker down if the future is so grim? The decivilizationists would see biding one's time as a cop-out, the pretense of doing something while actually doing nothing. But it's not doing nothing. It's fostering resilience. It's fostering *survivance*. Gerald Vizenor introduced that evocative term to describe what North American aboriginal peoples, even as their bodies and cultures were subjected to genocidal forces, always managed to do: survive, endure, and resist. Survivance, he writes, is "an active sense of presence over absence, deracination, and oblivion" (1). It's a refusal to be a victim even though you are a victim. It's holding on to what you've got even though everything is out of your hands.

Of course, I'm not entitled to appropriate this term—that would be yet another theft from those who have already suffered centuries of stolen lives and lifeways. Nevertheless, we'll need to channel something like the fortitude and moxie of survivance as we plunge deeper into this terrible century. We'll need the determination to persist and abide. And we'll need to prepare ourselves—without knowing exactly for what. Perhaps for something the Greeks called *kairos*: the moment when time and place bend together to form a brief window of possibility. Possibility of . . . ? Again, we don't know for sure. Maybe the chance to step out of the uncharmed circle of modern civilization. To throw sand in the gears and slip through the auger without being ground up. To find ourselves in a less unhappy place under a wide blue sky, blinking with amazement that we've somehow crawled out of a very deep and dark well.

Photo by author

iPod and World System

> People who are staring at their cell phones never rampage. They merely bump into things.
> —Edward Mendelson

THE AGE OF GREAT DISTRACTION

As glaciers give ground and beehives empty out, we are reminded that talk is cheap. The reefs blanch, the pines lean, and the algae blooms, but phone plans have never been more affordable. We can tweet out the end of the world for next to nothing.

There was a time when telephony was an occasion. Even now there remain older folk who, toward the end of a monthly check-in by a distant loved one, can be heard to say, "I should let you go now; this must be costing you a fortune!" These are the same people who resist the introduction of smart devices into the classroom, boardroom, and bedroom. They frown when their interlocutors brazenly toggle out of a conversation to answer messages or check scores. They recall the era when lunch meetings were possible without the intrusion of Facebook Messenger; when lighting a cigarette, not thumbing a scroll bar, was what you did with your hands when you were nervous or bored. Generally, they refuse to believe that the far is more compelling than the near, that the distant emailer should command more attention than the friend at the elbow. They are the ones who secretly wish to scream, "Asshole! I hauled myself across town to talk to you, but you keep jawing with them mopes what stayed home!"

In the age of ubiquitous communication, someone has hit nature's mute button. The earth is wounded, but its cries don't sound in the blabosphere. Ecological facts are mere tidbits in the info-stream. They are coded, slotted, and chewed through the appropriate media until they lose their flavor, drop from conversation, and harden like gum on the sidewalk. The stark science of a record flood or drought must give way quickly to the human-interest angle, whereby ratings share rises with the report of a drowned pet or a quick-thinking mom but not 400 ppm. More and more of us have something to say about the weather, but more and more no one cares. Whether you are a climate hawk or climate ostrich, there is only limited bandwidth for your squawkings. Too many other things to talk and text: which celebrity got busted; who's got a sale on mattresses; where's the best road-dog in town? Think you can sell global warming with a hockey stick graph or a picture of a swollen river? Forget it, Jake, it's Humantown.

THAT PERIOD OF TIME IN WHICH OUR AFFAIRS PROSPER, OUR FRIENDS ARE TRUE, AND OUR HAPPINESS IS ASSURED

In the 1960s, the future became an academic subject. Aided by popular treatments like Herman Kahn's *The Year 2000* (1967) and Alvin Toffler's *Future Shock* (1970), future studies specialists launched journals, institutes, and colloquia to explore the potential of the discipline. The result was, yes, not unexpectedly, the future was worth studying; it could be measured; there were features in the present that could signpost what was ahead; scenarios could be unspooled; the great question—"if this goes on . . . ?"—could be answered, provisionally, with the understanding that while history doesn't always proceed as expected, expectations can be tweaked and rejigged as time marches on so as to narrow in on the target. Today academic prognostication goes by the name of "strategic foresight" and "forecasting," and it has settled into graduate programs in business and professional schools. Back in the 1960s no futurist dreamed how much the rise of computers and associated information technologies would change the nature and comportment of the human animal. But they can dream now. The new infosphere joins the lithosphere, hydrosphere, biosphere, and atmosphere as those components of the earth system that define the contours of human life. The question we should ask scholars of the future and academicians of foresight is simple: Will the media ecologies we are cocooning ourselves within mitigate the encompassing eco-crisis or merely blind and deafen us to it further?

AMBIENT INTELLIGENCE

In *The Book of Fables*, the poet W. S. Merwin writes of the "Remembering Machines of Tomorrow." He observes wryly that the "human memory is a wonderful development but its fallibility is infinite. How can it be left to men?" (104). In short, it cannot. So man invents devices to keep track of what is too easily forgotten, from the "first notches bruised into bark" to "the air-conditioned archives of the age of history." The relinquishment of his memory is not unprecedented, for he "gave up his legs for the wheel. He gave up the strength of his arms for the lever. He gave up power after power of his physical form. And now at last, as more and more was forgotten, he began to relinquish memory so that something would be remembered" (105). Merwin imagines a time when each man will have his own personal memory machine that travels with him everywhere, recording his life as it proceeds. "The machines will retain, in flawless preservation . . . not only what their owners experience but what their owners think they have experienced, and will sort out the one from the other." At some point, it will be noticed that "experience is not only flowing into the machines but that they, to an increasing extent, are becoming its source as well. Man's experience of the mechanized memory of his experience—that is what will fill more and more of his days on earth. . . . His attachment to it will constitute the whole of his present—or what he takes to be the present" (106). It is not clear why this piece is called a fable or why its dateline is tomorrow.

BIG SHOES TO FILL

With all this data flowing here and there, all the remote sensing and satellite scanning, all the lenses poking into every nook and cranny of the planet, it's a wonder more people aren't less ignorant about where their world is headed. It's a puzzler. There are petabytes of information that address our down-spiraling ways, but the interconnectedness of the globe hasn't meant that there's greater concern about its looming collapse. The people ringing you from Bangalore and Mumbai just want to know if your ducts need cleaning. The rise of the datasphere has mostly blessed us with more gossip and porn. Globalization just signifies the usual busyness at a more rapid pace and in greater volumes: shipping of cars and corn; transmission of disease and Hollywood; shuttling of tourists and sales reps. But you wouldn't know, now that the world is said to be flat once again, that we are about to run out of everything: water, soil, energy, air, peace, and time.

The crux of the matter is this: As the world has become smaller, individuals have become bigger. The constraints on the growth of the imperial self have been taken off, and each person's ecological footprint is potentially size 13. The ecosphere is everybody's doormat.

What to do? Education, services, information technologies: we've been hearing for some time that in the frictionless economy these sectors will replace steelmaking, raw materials processing, and so on. The thought is that bad growth can be replaced by good growth, by exchanges that ephemeralize the means of production. E-books, for example, are already replacing paper books, wastefully pressed from precious forests. But we have to be willing to take this process to its logical conclusion. And what is that conclusion? Well, when it comes to automobiles we must ephemeralize this dirty mode of transportation into, say, the bicycle or the horse. How about giant fishing trawlers? Ephemeralized to a worm on a hook. The list is endless. Sixty-inch, energy-intensive LED screens could dissolve into storytelling circles or games of pinochle. What about giant factory farms and their torrents of blood and ordure? Presto! Say hello to the chicken coop and the manure pile.

Perhaps what we really need is a Great Leap Backward.

WILD KINGDOM

Few would be shocked if it were discovered that back in the day tele-zookeeper Marlin Perkins had staged a dust-up between, say, a leopard and an electric eel. Does anyone still care that Walt Disney stampeded lemmings over a cliff or slipped a peanut butter jar on a polecat's head to gain audience share on Sunday night in 1972? The classic wildlife filmmakers made animal problems look like human problems. But those old shows did more good than harm: By making animal lives into little dramas and picaresques, they got people thinking good thoughts about the critters. Even David Attenborough, who never faked anything, used his powerful hi-def cameras to render nature sleek and seductive. And a good thing, too. Watch real nature for any length of time and mostly what you see is nothing at all. Boring. For nature to be compelling, it needs a good editor. It needs an angle. The animals have to show some *motivation*; they have to *connect* with the viewers. "Brilliant footage, Clive, but where's the human interest?"

Case in point was the late Steve Irwin, who notoriously held his infant while feeding chickens to a saltwater crocodile. Steve was simply trying to vamp up the wild and splash biophilia across the screen. He made us think warmly about those wild animals he was romancing. This is good, right? Not so fast. The glamorization of nature through the various animal channels is the multimedia version of late nineteenth-century zoos, which epitomized in their urban containment of representative species the greater national containment of wildness, which would henceforth be found only in the deep woods or over the mountain, yonder past the last outpost. And today? The containment is complete, with even uncaged wildlife zooified as a result of the remarkable tools available to observe and display them. They cannot hide from our gaze; the cameras track them everywhere, into their burrows and undersea grottoes. Forget the animal porn genre—super slo-mo footage of stalking, killing, and gorging—because even family friendly nature films like *Free Willy* or *Spirit: Stallion of the Cimarron* promulgate the BDSM fantasies they fervently wish to deny: wild animals as kept women, naughty children, savage royalty. All but the most uncharismatic creatures have their media niche, branding scheme, and associated websites and plush toys. Where is the wild kingdom? Where is the beast that roams free of our eyes and escapes our visual pleasure? Nowhere. The whole world is one vast heavy-petting zoo, surveilled and stroked by a teledildonics of extraordinary invasiveness.

LET'S GET IT ON

We are led to believe our personal digital technologies are hooking us up with the incipient world-mesh. Our phones penetrate the clouds; data slide up from the ego to the planetary mind and back down again along invisible threads. It is one world, and we are pulsating nodes in the grand design. Oh, how we dress up our motives in fine universal cloth! To a man glued to streaming porn the information commons is just a glorified skin-book; the matter at hand is masturbating, not communing with the global village. We'll know soon enough if our talking planet can talk itself out of its continual self-abuse.

A GALAXY IN EVERY POT

The smartphone arrived none too soon on the planetary scene. Humankind needed a way to distract itself from aggravating reality, which was sending distress signals about fractured polities, soaring inequalities, and ecological calamities. Humans are the talking species, and what smartphones put into their hands was the means to communicate over all distances and times with equal sincerity. With a rapidity never before observed in the history of technology, there came a darn-near instantaneous blending of means and ends, instrument and purpose, so that in one fell swoop it became obvious that the point of talking was not the dissemination of messages or information but simply the act of communicating itself. To talk, to text, to tweet, to transmit: All that mattered was the capacity to ceaselessly signal "I am here!" to one's peers. "We all of us love to broadcast, to call ourselves into existence against the obliterating silence that otherwise dominates so much of our lives" (Harris 68). Thumbing out messages and checking for incoming seemed to be what progress had been leading to. We had arrived. The future stretched before us, and it was one vast chatroom.

What we've learned about ourselves thanks to smartphones is that the human brain does not wish to be alone with itself. The moments of solitude that we've been forced to endure in all our previous eras on Earth are now seen for what they are: unwanted downtime. The individual brain, previously thought to be the residence of entelechy, cogito, soul, what have you, turns out to have been a router that just needed the right kind of network connection to function properly.

ON A PERSONAL NOTE

When I was a grumpy grad student back in the last century (must have been the year we finally got serious about climate change, say 1992) and teaching yet another deadly section of what was known as "writing for people who hate writing," that is, freshmen comp, some whippersnapper pulls a walkie-talkie out of his knapsack and starts yakking during my lecture. Now there were two things wrong with this picture: one, you *never* lecture on writing, that's a means/end contradiction; and, two, unless you're calling in an artillery strike during a zombie apocalypse, no one, I mean no one, uses walkie-talkies in my class. Not. On. My. Watch. Heck, my walkie-talkie policy was right on the syllabus.

After some necessary roughness, some back and forth during which time the student got the lowdown on Triple C (Classroom Comportment & Civility), I learned what that walkie-talkie really was. Well, as you can guess, it turned out this incident was merely the first shot in what for me would become a twenty-five-year battle. My Existential War Against Constant Cellphone Intrusion (MEWACCI). We had gone down the rabbit hole, people, and what we found there weren't real purty.

What an age that was! The epoch when you could be in the middle of a nice meal with your wife in a fancy restaurant, talking smack about the latest Hendrik Hertzberg column in the *New Yorker*—she says, didn't he skewer Bush the Younger? and you're saying, Righteously, and she says, Whassup with Laura? she's a librarian, doggonit, and you, So they tell me, darlin'—and then, out of the lambent night, like the breaking of the seven seals, the guy at the table behind you screams, "YELLO? OH HEY SMITTY! YO-YO-YO! I'M AT SARDO'S. THE VEAL. FRIGGIN' TERRIBLE! WHEN'S OUR TEE TIME!"

Or you're in the supermarket, quietly takin' care of your grocery bidness like a champ, inspecting yogurt labels for best before dates and comparing milk fat percentages, when some bustling renaissance gal parks herself and cart in front of the Dannon and proceeds to get into a heavy cell discussion with her dispatcher: "Yeah, I'm in the dairy aisle now. Do we need any cream? I'm right there. Uh-huh. Uh-huh. Half and half or table? Or 5%? Or whole milk? Or Coffee Mate? They've got that new Amaretto flavor you like. Or buttermilk? Or goat milk? Or aphid milk? Or milk of magnesia? Or the milk of human kindness? Or arsenic? I like that. Uh-huh. We need sour cream for the potatoes. Regular fat, hi fat, lo fat, or no fat? Extra fat? What's that? Uh-huh. Uh-huh. Oh, you're definitely doin' smoked spuds in the Webber, hon. Remember how they were soooo good in the lava vent on Hawaii? With Duff and Tawny-Lou? Remember how they fell into that tar pit? How we left them there and went back for happy hour? Uh-huh. No, why would I make a grocery list? Why would I do that when I can phone you from the Piggly Wiggly and let the rest of the plebes get in on this exquisite display of impeccable taste I'm putting out here? Uh-huh? You say I could be more discreet that way and not annoy the hell out of all sane people who are needless to say totally uninterested in what we shovel into our pie holes not to mention the grey waste spaces that form our routine thoughts and feelings? Well, let me get back to you on that one, lover, I've got to take another call."

So we suffered through that era (OK, we're still in it). But it seemed as if some kind of politeness protocol eventually had taken hold after the years of early adopter anarchy, thought I, because there weren't so many ringtones going off in my classrooms like microstrokes in my head. But then less than a decade into the millennium I noticed another tectonic shift, and it was worse, and it was unstoppable, the Pacific Plate of sumbitch technologies. Inconsequential, addictive, primate-social-grooming-equivalent communication, you have a name: TXT MSG.

ALONE AND IN BAD COMPANY

You're starting to grok what I'm on about, right? Cell phones were just the thin edge of the wedge. A new public/private, I/Thou split was opening up—or maybe some old one was knitting over. Kinda hard to say which. The mind was extending its feelers into fresh niches even as stale ones were being bricked over, sealing off casks of fine mental amontillado, for the love of God! Your seat of consciousness was turning into an elaborate piece of sectional furniture and there were other people sitting on it. As MIT media theorist Sherry Turkle says, "alone together" is our new communal posture. Turkle had drunk the Kool-Aid when the online experience came along back in the 1990s, going gaga over life on and in the screen. The forging of new identities, new communities, the liberating potential of cyberspace, and all that jazz: Sherry was pumped. Now, twenty years on in this brave new world, she has belatedly figured out what anybody who was paying attention already knew: that life, love, and interconnection carried out in virtual worlds takes an awful lot of time away from their real-world equivalents. All this tweeting and chatting and snapping and pinning and liking, well, the hours go by, don't they? As Marshall McLuhan explained, it's not the content of a medium that matters but the medium itself, what it does to your body and mind, your social forms and norms. If you want a good image of how this works, just think about a kid spending eleven hours a day playing video games. Doesn't matter what the video game is: you name it, Super Mario or Grand Theft Auto. The point is he's down in the basement in the dark moving a blob around on a screen; he's not outside catching frogs or playing scrub with his pals. His mind is being formatted for computer logic trees; his body is being shaped for Swedish office chairs. A few years of this and you produce a creature well-suited to his future career as a symbol manipulator in the digital pin factory.

As Turkle learned on the road back from Damascus, it seems as if the call of the remote and the virtual too often trumps the immediate, shared, lived experience. Texting is molding us into arms-length companions; our relationships are becoming superficial, loosely connected, and easily trashed. When the in-box dings, the devotee of the Church of BBM responds like Pavlov's dog: father, mother, sister, brother, friend, colleague, boss, underling, all go under the bus as the pod-person reaches out and swipes in the message from afar. The text chatter begins, and the residual f2f devolves into grunts and false affirmations: "Uh-huh?" "Yep." "Is that right?" "Sounds interesting." That's what we call *phatic* communication, basically just noises you make with your mouth to signal to the person you're nominally with that you still have a pulse. Meanwhile, you're busily typing the extremely urgent communiqué back to the remote interlocutor, who by virtue of his nonpresence is somehow more fascinating than the living being across from you: "I am @ lnch. Lets meat @ 2 @ Starbuks 4 coffee. How about them Cubbies?" And when you hie thee over to Starbucks at two, you will in turn field all the texts you're getting from work as you fake a new unengaging conversation with the latest warm body, who is anyway multitasking on his own Android tricorder. WTF is with us? A sad truth: in flesh, we are no longer very compelling. But as distant and truncated texters we become lively and engaging entities, who can charm and delight with our out-of-the-ether apothegms and koans. LOL.

Now, you might argue that what alone-together really means is that there was already a huge hole in our so-called holy communion. Why else would we so quickly attend to the far if the near was fine and full in the first place? The text rushes into that ragged spirit-opening like a tornado into a low-pressure zone in Kansas. As we arrange our bodies across a table, what we imagined was to be a meeting of minds turns out to be two hunks of Swiss cheese passing in the moonless night. Text messages, in this view, make up for the fundamental poverty of interpersonal communication; they are the dangerous supplement that is not really a supplement at all but a welcome relief from the meagerness of our mostly vacuous common experience.

I hear you. (I'm actually listening to you; I'm not checking my email. Honest.) And I know you are world-weary and you've seen it all before and you're also bored and you prefer grins to grimaces and you don't need me in your ear and you're receiving a text anyway. You have teenagers; you have a busy life; there is stuff U 1/2 2 DO! The world is demanding; it claims your attention. You say that texting is wonderfully efficient. It lets you control the place and pace of your communications. It gets you quickly to the gist of things, to those hard kernels that vibrate with cosmic significance: "When r u picking me up?" "Check out this cat on You Tube." "I'm sew board." Moreover, you tell me that people have always griped about new modes of communication. You say that we never know for sure how technology is going to be used in advance, so it's premature and plain dumb to get riled up before things fully play out. Water finds its own level; texting will, too. It is the circle of life, Simba, relax and eat your antelope.

Tools get proved on the ground, you philosophize, your thoughts growing expansive as you put aside your phone and remember how to construct an argument. Humans, you observe, will use tools in ways that suit their needs. That's Progress . . . nay, that's Evolution! We are as the trees and the stars, and we have a right to be—and to tweet! You can't stop the future from coming! Get on the train or get out of the way; no matter what, tomorrow won't wait! And it's more wonderfully strange than you in your flimsy house of cynicism can dream of! Why, didn't Thomas Edison invent the gramophone in the 1870s because he thought people would need to record telephone conversations? Gee, you say, getting all sarcastic, what would happen if the farmer's wife down the road on the party line gave her friend a recipe for shoofly pie but the friend forgot the list of ingredients? Thanks, Thomas, for this wax cylinder that helpfully recorded those evanescent words for all time. Ah, yes, here it is: two gills of blackstrap molasses.

C'mon, you explain (getting all indignant and professorial on me, impatient with my dopey Luddism), we text because it serves a need, a lack; it improves our lives, for crying out loud! You can fight reality all you want, but folks vote with their feet—and their fingers. Don't mistake your heat for our light, Ned, thinking you can scold us into ditching our tools or using them according to some fussy etiquette. That's for History to determine. As Eddie Murphy told us back when he was still amusing, "White men stole George Washington Carver's recipe for peanut butter, copyrighted it, and reaped untold fortunes from it, while Dr. Carver died penniless and insane, still trying to play a phonograph record with a peanut." "Q.E.D., blockhead!" you shout. 'Nuff said.

OK. You've made your point, dear reader. I'm in the way of techno-destiny. I get it. But relent a bit, willya? Can I respond? Thanks. Now, I don't exactly understand what Eddie Murphy has to do with this, but I appreciate your vigor if not your rigor. You speak with the passion of the convert. But, alas, you couldn't be more wrong.

You've made the error that all junkies make, and that is to confuse what you love with what is good for you. A lot of people love butter though it hardens their arteries. A lot of people love Tom Cruise though he hastens dementia. And a lot people love texting though it contributes to the demise of civility and the rise of inattention. Yet all too quickly we forgive the things we love their unforgivable faults. In fact, we often don't see them as even possessing faults. We adore our ugly babies. Maybe love is really just lack of taste elevated to virtue. Keep in mind that humanity is a self-flattering species: We have a hard time looking in the mirror and not seeing the fairest of them all. And we gaze down at what we have wrought and see that it is good, darn good. How could it not be? After all, it's a little piece of us.

And Man, looking on the smartphone, saw that it was totally awesome.

Here's what you must do to help save civilization from itself. The next time you are having a nice talk with your friend Joyce, and she even so much as flinches when her mobile device buzzes her about an incoming missive, politely rip the device from her hand, place it carefully on the ground, slide the ball-peen hammer from your purse where you previously stowed it, take four or five practice swings, then with all the main force you can muster, obliterate this devil's device into its component molecules. Don't worry about Joyce's reaction; just tell her it's for her own good and that eventually she'll understand. (The thing was probably free anyway—-they get you on the plan minutes.)

Then say, "Now Joyce, where were we? I think you were telling me about Bob's new grapefruit diet . . ."

Got it? Can you make this happen for me? All social movements begin with a single, heartfelt, "No more!" I'm counting on you. And think of what awaits you. Now you and Joyce can finally have a nice chat in the here and now. Won't you feel better? And, thanks to you, won't we, all of us, be just a little closer to finding our humanity again?

INVASION OF THE MIND SNATCHERS

As humans sink deeper into their pods, critics are hard-pressed to square the new electronic conviviality with folks' general blindness to the disintegration of the world systems. Wasn't this to be an era of unparalleled awareness? These hand-caressed devices, tethered to a worldwide information net, can summon knowledge of the state and velocity of our planetary conditions with the precision of Maxwell's demon. It's a safe bet that if one wanted to know, say, the probable decade when Phoenix will become unlivable, a ten-year-old could find out in thirty seconds. But few want to ask that question, let alone hear the answer—especially those living in Phoenix. On the other hand, millions follow via Twitter the daily epistles of Katy Perry, Cristiano Renaldo, and Kim Kardashian. Our taste for information runs toward the soft and oleaginous; the rest of it can go into the slop bucket. It's as if a starving man had been turned loose in a Golden Corral but all he ate were the butter pats.

There is a theory that "information overload" explains the preference. What Avital Ronell said about the telephone seems even more apt for the smartphone: "When you hang up, it does not disappear but goes into remission. . . . There is no off switch to the technological" (xv). Bombarded with grim news of the world's shrinking prospects, we shelter in the more human-scaled mishaps of celebrities, the latest gossip from around town, and the jabbering of friends and neighbors. The big picture stuff is swiped off the iScreen in an iHurry. So it's not as if people don't know there are problems with the climate and all that, it's just that they can't process the garish details. Far more efficient to move on to details they can process. "Poles are melting, yup, but tonight's half-price pitchers with a pound of wings!"

But this is a trite answer, almost a truism. People have never liked bad news, that's for sure. The more intriguing possibility, for critics of the datasphere, is that our little devices are actually working to compress and diminish planetary-scale problems. The medium miniaturizes the message. Smartphones are potent machines that put the universe in one's hand. The dedicated user must feel a sense of invulnerability and control, like James Bond with his Walther PPK. The more we port our lives and linkages through these elegant gizmos the more we feel empowered to shunt the world's woes into a mental black box. It's a reverse Pandora's jar, with all the mischiefs recontained and tamped down. We can literally keep the dismal information off the screen and out of mind. If you're like me you rarely empty your vacuum cleaner, so there in the closet, safely bagged, is all the filth you've removed from your house over an embarrassingly long stretch. The mental hygiene performed by the smartphone works on a similar principle, sucking the soiled goods of the media environment into a neat, clean package. If you want to deal with it, well, that's your business. But if not, let it lie dormant in the dark.

ANTISINGULARITY

Still and all, what a great time to be alive! Or, perhaps, better to say, what a great time to be alive if you can get your life in under the wire—say by midcentury. After that we're not so sure. But it's amazing how everything is coming together all at once: dream technologies, like the iWatch and the bionic eye; and ecological nightmares, like the extinction of bluefin tuna and monarch butterflies. You've got on the one hand self-driving cars and on the other climate change drivers. You have 3-D printed guns and methane clathrate guns. Cadillac Escalades and trophic cascades. At the high end, we see globalization uniting the world through trade, and CO_2 disuniting the world through lax emissions treaties.

The science fiction writer Vernor Vinge popularized the concept of the singularity, which seems to owe its origin to an exchange described by Stanislaw Ulam in his eulogy to the polymath John von Neumann: "One conversation [with Von Neumann] centered on the ever accelerating progress of technology and changes in the mode of human life, which gives the appearance of approaching some essential singularity in the history of the race beyond which human affairs, as we know them, could not continue." Vinge argued the singularity Von Neumann identified would consist of an extraordinary explosion in earth-based intelligence, including "superhumanly" intelligent computers and networks, "intimate" interfaces between machine and humans, and biologically enhanced human intellect. These concepts were later explored by Raymond Kurzweil in *The Singularity Is Near*. He predicted an unstoppable acceleration of intelligence, one that "will continue until the entire universe is at our fingertips" (487). The nature of exponential growth has always been hard for the human brain to process, and perhaps as a result it took Kurzweil himself almost five decades of immersion in computers and information technology to finally "become aware of a transforming event looming in the first half of the twenty-first century" (7). *The law of accelerating returns*, Kurzweil explained, is a stealth law, whose full meaning is not clear until it is in full swing. Just as a single penny doubled each day for a month yields modest sums for three weeks until it mushrooms to over five million dollars on the thirtieth day, incremental increases in knowledge will suddenly produce scientific and technological leaps of breathtaking scope. Massively multiplied mind power, according to Kurzweil, will be particularly dramatic in the areas of GNR: genomics, nanotechnology, and robotics. (The genetically enhanced lichen that assists in adding volume to boreal trees—thus locking up gigatons of carbon dioxide—in science fiction writer Kim Stanley Robinson's Science in the Capital global warming trilogy is precisely the sort of game changer Singularitarians believe is in the pipeline.)

If the Singularity really is near, we can stop worrying about a number of currently worrisome matters: population explosion, resource depletion, ecological collapse, species extinction, food production bottlenecks, and so forth. The Singularity will reduce these problems to chump change. If, on the other hand, the Singularity is just another verse in the tired old song of human overreach, a different sort of singularity—one we can actually confirm, not speculate about—will soon be on us. Let's call it the *Antisingularity*. It is a convergence that obeys *the law of accelerating calamities*. The double-edged sword that is instrumental reason leaves no mark on Singularitarians like Kurzweil or Vinge; nevertheless, its cutting logic is well-understood. The whole world, including man himself, becomes standing reserve, to be used as reason sees fit, to paraphrase Martin Heidegger. In fact, we are already finding that redounding environmental and social stresses are undermining, not enhancing, the capacity of human intelligence to deal with the impending planetary breakdown. And far from being swept away by rising knowledge levels, the Antisingularity is in part produced by them.

What Kurzweil and other true believers have forgotten is that always in history an expansion of human possibilities has meant impossibilities elsewhere on the planet. Progress goes hand in hand with regress; the shiny new highway drains the old dismal swamp. And now all those swamp accounts are coming due. Intelligence of a different sort needs to be applied during the Antisingularity—not superhumanly artificial intelligence and cyborg implants but something the ancients knew as *phronesis*, which is to say, practical wisdom and prudence. Soaring smarts and transcendence dreams count for nothing without the plain-speaking that says, with Mary Poppins, enough is as good as a feast. Bernard Stiegler observes, "We know that in the coming decades, the Earth and her inhabitants, human beings, will have to demonstrate like never before—individually and collectively—the worldly intelligence and sense of responsibility that, in principal, define them as human beings rather than cruel, vulgar, and gluttonous slugs" (3).

FOR A NEW PLEISTOCENE!

Perhaps with great intelligence we will discover the way forward is to turn back the clock—turn it way, way back. Let's think about this: Information technology is liberating us from the drudgery and unnaturalness of the industrial machine, automating the nasty business of making a living. Soon we might be able to plunge through the looking glass of ubiquitous computing, offload our workaday burdens to bots and droids, and enter a new space of sensuous freedom wherein we can unlimber our primal selves and reestablish our lost connections to so-called reality and much-missed authenticity. We're all hungry for it. In this postdystopian cybertopia, when folks are on vacation at the beach, you will not see them checking their phones for incoming missives. The TV screens that line the walls at sports bars will go unwatched as patrons sink into the warm embrace of primate sociability. We will retreat from the media room to the Zen garden next to the backyard koi pond. Teenagers will allow their device batteries to run dry and never bother to recharge them. We will rediscover our true selves as Edenic beings built for joy and pleasure. The day will come when we all be out in the woods running around naked with mud on our faces.

ENTER THE MATRIX

Or perhaps information technology will foster a new inner wilderness, replacing the outer one that IT is helping render uninhabitable. The age of man qua man will end; then comes the time of man-plus. Unaugmented man was the problem solver; he was always looking for things to fix. He needed fixes because he was forever breaking things in a world still bound by nature's laws. But man-plus is no fixer in that sense; he will tarry in the chaos of the digital wilds, following his inner lights, every day encountering things gone awry but thrilled to create workarounds, for this wilderness obeys *his* laws. He will race past outages and skirt glitches. Triumph, fail, eat, or be eaten: it's all of a piece. No plan, no purpose, nothing but unfettered play in the constant buzz and bash of virtual facticity. Man-plus will be a strangely feral sort of creature, rewilded by design, inhabiter of open worlds of his own making. Outside his domes and subterranean crèches, a grim, depauperate planet will go about its business of finding a new equilibrium. Inside, man-plus will sink ever deeper into the playground of the mind.

Photo by author

Entertaining Futility

This futility had let's call it a flavor.
—Jonathan Franzen, *The Corrections*

WEASEL KNOCKS OUT PARTICLE ACCELERATOR
Please excuse me [she began] for adoring films in which animals abuse humankind. What an instructive genre! It's wish fulfillment in cinemascope. Think *The Birds*, *Planet of the Apes*, *Day of the Animals*, *Zoo*. In each case nonhumans visit a richly deserved nemesis on the man-made world (frequently embodied in the form of B-list actors), bringing about not only its destruction but also—and this is my opinion—a salutary if rather painful transition to a postanthropocentric existence. My theory is that all such fantasy is about, ironically, the promise of permaculture. When Tippi Hedren and Rod Taylor tiptoe into that bird-infested landscape at the denouement of Hitchcock's greatest film, audiences wonder if the nightmare is ending or just beginning. My vote is for a new solidarity between nature and humans. (Tippi, by the way, went on to become an ardent animal rights activist; Rod later cameoed in *Kaw*, a made-for-television homage to the original avian classic.) When animals attack: such films are rarely just about claws versus smarts. Who can forget the menacing dugong in the Australian cautionary tale *Long Weekend*? I admit you must dig deep to extract the moral lesson. Yes, the film focused on the knee-jerk ecophobia of an unlovable bickering couple played by John Hargreaves and Briony Behets, but you don't need a cast of thousands to drive home the point: man proposes, nature disposes. Remember *The Andromeda Strain*? A satellite bearing extraterrestrial microor-

ganisms crash lands in a small New Mexico town. James Olson and Arthur Hill investigate. Only a colicky baby and a Sterno-drinking old coot manage to survive the outbreak. Returning to their high-tech underground lab, the scientists determine the cure lies in producing an abnormal blood pH. Unfortunately, the inventive germs escape containment. They learn to eat plastic. Supporting actor David Wayne is trapped. The white coats must violate all the facility's built-in safeguards to outfox not only the germs but also the nuclear self-destruct protocols put into place by a paranoid military-industrial complex. Thank goodness: turns out nukes would only have strengthened the atomic energy-loving superbug. Well, big sigh; the world is saved. Nevertheless, it's implied the germs will mutate again. To survive, humans must transcend their current state of self-absorption. *Quelle surprise*! Then there was *Phase IV*, helmed by graphic designer and film title sequencer Saul Bass. This was more proof of concept than movie. The concept is that humans are at a developmental standstill. They are destroying the world and can't help themselves. (Sound familiar?) It takes another cosmic event to set things right by turning ants into our overlords. The surreal ending that Paramount cut for no obvious reason popped up a few years ago. In it Bass has Michael Murphy and Lynne Frederick running around like lab rats in giant mazy geometries created by the hyperintelligent little pismires, who are beneficently seeking to move us up the evolutionary ladder. Dialectics once again. I have this little screenplay in the back of my head that would take the genre toward its logical conclusion. It's about meat animals turning the tables by placing *us* in the high-density feed lots, the poultry warehouses, and the swine barns. Kevin Bacon and Megan Fox would star. We're the ones neutered and defanged, our ears cropped and our fingers snipped so we can't tear at one another. On a high-protein diet our muscles soften and marble. Of course, there's going to be a plot: a few of us manage to escape confinement, elude the posses of dog-riding rabbits, and light out for the Forbidden Lands. We manage to put together a little rebellion, Spartacus-style. For a time, things are looking up. Action, romance, tears, then more action and more romance. But, sorry to say, we are beaten down in the end. Nevertheless, our unexpected uprising has threatened the foundations of animal society; it becomes, if you will, bovo-polo-porcocentrism's Copernican moment. In the beast high-counsel chambers, the question is posed: should humans have rights? An issue to be debated for centuries. Some day, my screenplay suggests, humans might be let out of the pen. Is my film a metaphor? I'll leave that for you to decide.

STARS' HOMES THREATENED

The news out of Topanga Canyon wasn't good. Fire was racing over the chaparral. At the avocado ranch, fruit was exploding like hand grenades. Reality TV celeb Tiffani loaded up the jeep but Kisses the bulldog was missing. The film and other creative industries account for over 3 percent of the US GDP. Firefighters called in the water-bombers. Some crusty old dude was still refusing to abandon his bungalow; he kept spraying down the roof until the hose started melting. The governor declared a state of emergency, but that didn't slow the fire a lick. One editorial asked the question, is global warming to blame? Another said this was no time to politicize the tragedy. Slash and burn season in Borneo had begun. Orangutans swung through the smoky trees as eyes watered in Jakarta. Palm oil represents more than 10 percent of Indonesia's export economy. This saturated fat has the enviable property of remaining semisolid at room temperature. In Fort McMurray massive wildfires incinerated one-tenth of the buildings. Eighty thousand townsfolk skedaddled. The nearby oil sands projects were shut down indefinitely, and a million barrels a day of synthetic crude stopped flowing. The interruption hit Alberta's economy hard, which had charged royalties of $9 billion the year before but now expected to collect less than $2 billion. The coal seam fire beneath New South Wales' Mount Wingen continued to burn. It had been smoldering for six millennia, moving at about three feet per year in the direction of Newcastle, the world's largest coal port. Australia sends 40 percent of its bituminous to energy-poor Japan. Back in the Hollywood Hills, Leonardo DiCaprio looked down over his city, then flew to Brazil for the World Cup. He bunked on the 470-foot super yacht of Sheik Mansour bin Zayed Al Nahyan, the deputy prime minister of the United Arab Emirates, head of the International Petroleum Investment Company, and owner of Manchester City Football Club. The sheik is also an accomplished equestrian. Mr. DiCaprio's first motion picture appearance was in *Critters 3*, in which he fought a band of charismatic alien carnivores.

WIN THE FUTURE

Innovation is the key. It's all about building value and controlling it. You're after wealth? It's in your hands—or rather your brains. It's not even brains: it's DNA. All humans are by nature creators and risk takers. Find and own a disruptive idea, then quit your job and sell your house, your car, and your firstborn to support that idea. Nurture it, grow it, invest in it, seek others to invest in it, venture capital will help, bring in brilliant collaborators who will live and die for the idea, get it to market, show it, brand it, push it, sell it. Get all over it like an elephant on a peanut tree. This is not your parents' world! It's not a world made of stone; it's a world made of water. Nothing is permanent. Like a wave, an idea forms only momentarily. You ride it 'til it breaks. Then you ditch and scan for another. Your parents wanted to live in castles—permanent and safe. But you're a surfer, a barbarian, a goddamn Tahitian. You're rolling out your own life, not following a path. What's that? You're wondering if maybe you'd prefer a roadmap? Maybe keeping to the highway is a safer bet? Uh huh. So you wanna stay in your lane and end up in a gloomy, dank pile of rocks praying for salvation? To the god of *Thank you, sir, may I have another*? Well, it's decision time. And by the way: all the castles are being pulled down. You'll live in a trailer park, figuratively and probably literally, that's your other option. Not much of an option. So grab your board and hit the beach. By our calculation, if we can squeeze out one awesome idea per person, that's seven billion awesome ideas. Do you think our planet will be worrying about poverty, crime, pollution, and disease with that kind of crowdsourcing? No way. That stuff is so gone. We'll be crushing it in paradise. We'll make the future we want and move into it. We'll all be Polynesians, living in grass huts and drinking out of coconuts with Ginger and Marianne. Pig roasts and fire walking! Weather? Fuck the weather! We'll *own* the weather. The only thing we'll have to fret about is this: steak or lobster? And how about both?

THERE IS NO REPRESENTATION

We have only statements and other statements, experiences and other experiences, feelings and other feelings, thoughts and other thoughts. The mind is associational. Signs lead to signs, not to things. The story of the three bears, the Yellowstone bears, and then the story of the bears again. A poem of Mont Blanc and Mont Blanc: two objects that make separate claims on us. We hear of a flood, a drought, a fire, and a wind. All are made into movies. You swim, thirst, blister, and blink. All are made into Facebook entries. An old man walks through the woods and thinks about the woods seventy years before. Then it was all thicket and underbrush. Now it's open and high and the birds flit about unhindered. There are no microcosms, nothing fits in a nutshell, and you can never get down to brass tacks. The world exceeds your stupid reductions. Metaphor lies. Simile lies. Analogy and parable lie. Even fable lies: The dog really did want the hay and the fox really didn't want the grapes. To glimpse the

Author family photo

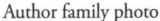

future, we read *Brave New World* and *The Handmaid's Tale*. But the future unfolds as it will and the novels unfold as Huxley and Atwood wrote them. Here is a picture at Lake Michigan from 1967; I am in the picture, with father and grandmother. But that is not Lake Michigan, and that is not me.

OLD DEEDS FOR OLD PEOPLE

I direct the students to self-organize into small groups and discuss the narrative's subplots, with a view to reporting back their findings to the larger class. Wandering around listlessly with my half-empty watering can of pointers, I notice the hand. A young woman begins, "Professor . . . Sorry, what's your name again?" I am reminded of my uselessness. "A month into this course and you still don't know who I am? Do you also not know the names of your other professors?" Probably should not have gone this route; no good can come of it. But I, Nemo, miff easily. She wilts, goes reflexively to her phone, always near to hand when not in hand. That's the safe zone, omphalos of spirit and joy. She smiles at the screen. Dismissed, I step off.

FOLLOW YOUR BLISS

When a human being is doing something with passion, he believes he has the world on a string and all the rest of us know it and are jealous. "Doesn't get any better than this, eh?" said the philatelist to his bored companion. Even the most run-of-the-mill activity performed by the humblest schmo gets the same kind of treatment from the inner cheerleader. I don't doubt that when the drywallers at the new Walgreens were really clicking they felt they were dining at God's table. Every author experiences these moments of swell-headed confidence: the notion that no other book is as vital and fascinating as the one she is working on. For writers, such cheek is probably necessary to get through the ordeal. Mark Twain understood well the self-absorption of the artist, but also how task arrogance is simply the manifestation of a wider species hubris: "'The noblest work of God?' Man. 'Who found it out?' Man." Arrogance, narcissism, egocentrism—designate this human flaw as you wish. We know it when we see it, and we deplore it when we detect it (in others). The problem is that we, unlike Twain, fail to spot the flaw as it manifests in its collective grotesqueness. As a species, we imagine we move through history and across the planet like a supermodel on a fashion runway, sashaying with purpose, imperturbable, chin held high. We own this place, doggone it, and the whole universe loves us. But surely the universe sees a different creature: the gibbering, black-souled ghoul, gnawing frenziedly on the planet's remains. This ineradicable vanity, this profound self-worth, seems like the ultimate driver of all human systems. Call it the will to survive, boil it down to selfish genes, and acknowledge that every living thing must possess it to some degree, but don't deny that humans have it in spades.

IT'S ALMOST LIKE SCIENCE FICTION AT THIS POINT

An article in the *News-Review* yesterday reported Greenland was warmer than New York City—about 36 degrees above normal, much of the ice sheet covered in a thin layer of meltwater. That reminds us of a day here in Pellston (a.k.a. "America's Icebox") back in March 2011. The previous all-time high was overshot by 32 degrees. It was about 85 or so. One of us remembers saying, *Have we entered the friggin' Twilight Zone*? Well, wherever we were, it was no place for making syrup. The sap wouldn't run. How are you going to collect sap without freezing nights? We've got 10,000 taps and nothing was flowing. Damned if it wasn't balmy at midnight. We couldn't sleep. We took off the storms and opened the windows to let in some air.

In twenty-four hours, we went from parkas to undershirts. The sap lines were drooping in the sun. About the only thing missing was black flies. Maybe it was too hot for them. Well, that's Michigan weather for you: if you don't like it now, wait five minutes. But global warming? We weren't convinced then and we aren't now. There's evidence on both sides. Some scientists believe we're headed into another ice age. Anyway, it's not like humans can truly bollix up the atmosphere. The sky's too big and folks are too small. Natural cycles explain a lot. The climate has always been changing. Still, if spring starts coming earlier, we'll just back up the syrup season a few weeks. And, honestly, it wouldn't hurt if it was a tad bit milder around here. You try living above the 45th parallel in January! Sure, we lost money that year, but then we made up for it the next. Besides, if the whole thing goes bust, we'll just lumber these trees and head to Florida. We're looking at a condo near Winter Park. We're not getting any younger. Here's an idea: a smart young person could think about growing maples in Greenland.

SINS OF THE FATHERS

The central event in Kurt Vonnegut's *Deadeye Dick* is the reckless discharge of a rifle by twelve-year-old Rudy Waltz. He fires it out an upper-floor window, aiming at nothing, a symbolic "farewell to my childhood and a confirmation of my manhood" (70). Soon afterward, he and his family learn the bullet has struck and killed a pregnant woman. Instantly, his father claims full responsibility, taking the fall for his young boy and spending the next two years in prison, effectively sacrificing his own future. "And never, may I say, would the moment come when he would give the tiniest crumb of guilt to me. The guilt was all his, and would remain entirely, exclusively his for the rest of his life" (77). Our world's plot is a reverse *Deadeye Dick*. In this story, the father annihilates future life by spraying machine-gun fire in all directions. Before justice arrives, he wipes his prints and places the gun in his son's hands.

FINITE JEST

Believing I hold no beliefs tightly, I do not know when I have arrived at certitude. To posit a belief in the first place means I am still questioning its validity. Whatever it is that is happening when I profess belief, it cannot be the articulation of granitic inner truth; it can be only an announcement about what my active mind is currently gravitating toward. Thus, conventionally defined belief is always malleable and impermanent in that it is continually tested through application and quite often found wanting. You say you believe in charity until it is asked of you; you say you believe in freedom until somebody else infringes on yours. These shallow beliefs are merely conjectural: motives draped over whims. By contrast the truer, profounder beliefs I wish to expose are buried so deeply no amount of digging will bring them to the surface. A belief of this sort would be what one hews to even as it is under constant pressure to be abandoned. One has so little conscious awareness of it that to abrogate it does not signify. Here is the conviction that steadies the hand of the suicide or heaves the soldier's body onto the unexploded grenade. No one suspects the existence of this staggering self-assurance—but, suddenly, there it is. Lying below the level of awareness, such obduracy is the unthought of thought, the blind spot of insight, the bedrock of the mind from which rise the primal discriminating capacities. Thus, true, unbending belief is practically no different from instinct. Now we must ask: Where are those true beliefs that would save us from the stupendously dangerous forces we are unleashing? Those beliefs akin to what prompts rodents to flee from fire or flood, waterfowl to migrate in advance of frigid temperatures, and sea slugs to hurl themselves into wild currents to escape predators? Do we not possess similar degree-zero commitments to survival? Have we not been equipped to answer decisively those threats self-produced on the grandest scale? Were we made to persist only to this moment of high reckoning and not beyond? Will the human species be extinguished for a lack of faith in its own necessity?

ALONE IN A WORLD OF WOUNDS

Dear Mom and Dad, they're making us write letters home this morning so here's mine. The wifi sucks anyway and they took our phones today so I can't text. It's supposed to be good for us to have a No Phone Day but everybody is freaking out. Anyway, it's OK up here. We are making something cool in the craft cabin that will be a surprise for you on Parents Day. The food is good but there's never enough of it. Just saying. Thanks for the care package especially the Oreos. There's a tuck shop but it's mostly Freezies and Popsicles. Yesterday we did water sports—canoeing, sailing, and swimming. We have to swim in the pool right now we're supposed to stay out of the lake cuz of the parasites especially the Gardia (sp?). One kid says there is also a Brain Eating Ameba which I think would be a great movie title. Oh, we played capture the flag after dinner. My cabin lost as usual. Why? Because we all suck at running. It's boring a lot of the time with the crappy wifi. There are like two cabins with good connections and the rest are slow as malasses. I'm not saying I'm not having fun but I don't see why they can't get some decent wifi up here. It'd be more fun if everybody could get online and play together. We could get to know each other better. I have made a sort of friend named Amir. We sit together in the food hall and play some bluetooth games. One thing we did that was cool was an overnight canoe trip down the lake. We were all really exausted after that. We saw some cool things on that trip such as huge cottages, powerboats, sea-do's, an abandend mine, and a pontoon plane. There is a lot of Algie in this lake. We had cookout and told creepy stories. Our counsler Blake said the guy who owns the plane was in a famous band. One time this guy flew in on the plane when he landed on the lake he drove right over a teenager in a kayak and killed him. It is said the ghost of the dead youth haunts the lakes murky depths. So we had to paddle superfast to get across. Everybody was freaking out and exausted! We went fishing a couple of times. There are supposed to be Trout in the lake but they weren't biting because they have to go down to the very bottom where there is cool water. One kid caught a gross fish called a Snakehead and Blake killed it by stabbing it in the eye with his swiss army knife. (Can I get one of those?) Everybody freaked out! On Saturday we went for a hike which was fun but also sad. The whole woods around the lake is pretty burned up from that fire last summer. The trees that are left are like black telephone poles. There is some grass and stuff that makes it a little green but it's mostly ugly and grey and you can still smell the smoke and ashes got all over our clothes. We didn't see any Birds or Animals which really sucked. I don't like to complain but it is boiling up here and the cabins have no air conditioning. I know this is supposed to be the north woods but it's weird with the woods being burned down and so hot and dusty all the time and the lake like a warm bath. When you come up Mom can you bring me stronger allergy pills? My eyes are itching like crazy. Blake says there is a lot more Ragweed nowadays than maybe when you were going to camp Dad. Another thing to tell you, sorry if it sounds like I'm always complaining, is that three kids had to go to the hospital to get shots. Blake says it's Lime

Ticks. Sorry also that my printing is so atroscious (sp?). Love, Eric

PASSENGER HUMANS

Great shoals of living flesh in the oceans of sky. Half a million people in flight at any given moment. They don't blot out the sun, but they do leave their mark. An empty 747 weighs over three-quarters of a million pounds. Add in 90,000 pounds of humans. The presidential version of this plane can fly a third of the way around the world without refueling. The Cessna 172 has been produced in greater numbers than any other airplane; it seats four. In 1958 such a vessel stayed in flight continuously for almost sixty-five days. Fuel and food were passed up to the plane from a convertible automobile. Virgin Galactic is working on suborbital flights; seven hundred future astronauts have already paid deposits. An average flight from New York to LA produces about sixty tons of carbon dioxide or about a quarter pound per passenger per mile. Thoreau wrote in his journal, "Thank God men cannot fly, and lay waste the sky as well as the earth. We are safe on that side for the present" (*Heart* 217). Forty-two years later Orville and Wilbur Wright made a sustained heavier-than-air flight.

THE WONDERS OF ANCIENT LYDIA

When it ended, I mean, it *ended*. Everything shut down. Shops closed. Schools let out. Factories shuttered. Buses quit busing. Service stations stopped serving. Even the mall locked up tight. "Police and essential services will continue," the radio said. But who knew what that meant out here in the 'burbs? We'd discovered we were miles from anything essential. Nobody had any gas; nobody had any juice. The racket from portable generators had stopped days ago. Around the crescent we just sat in lawn chairs waiting for the cavalry. Every few hours, Larry would yell over at us, "Hot enough for ya?" He was our neighborhood funnyman. Yeah, it was hot enough for us. The houses were pizza ovens, so a patch of shade under a tree was where you wanted to be. Some kids even got up in the branches, claimed it was cooler up there. At night, the blurry moon rose over the blacked-out city. Dust was filtering in from farm country. It was some sort of freak inversion. The air mostly just lay on the land like a wool blanket, occasionally tugged here and there by feeble breezes. A few folks hiked down to the river, but how was that going to help? The water felt like the air: hot and dry. Sometimes dead animals floated by. Sometimes? Should say all the time. It was like the damned Ganges. The fresh food was long gone. Bread, cheese, eggs, forget about that. Peanuts turned to peanut butter in your mouth. Pickles went limp. Crackers were impossible. Nothing crunched anymore. Even the patio tomatoes had gone to mush. Canned food was it. Ever eat a jar of warm jam? Not cordon bleu. There was slick of congealed ice cream on the floor under the fridge. How do you clean that up? The water in the bathtub was for drinking. Around the fifth day poor old Toby gave up the ghost. He'd panted so much there was no drool left in him. We buried him under the big spruce, his favorite rolling-around spot in the backyard. We could hear the Gordons over the fence talking about driving up north. But where could you go running on fumes? It was the same everywhere. The radio said be patient. At that point Bob Recoski started bitching about FEMA. Larry yelled, "Foolishly expecting meaningful assistance!" The whole experience was godawful, but maybe the worst part was the boredom. We hadn't been bored since high school. There'd always been plenty to do, what with work, kids, TV, home improvement, maybe the occasional booze-up. But now there was nothing, literally nothing, to fill the time with.

We just sat there and fanned ourselves. Then a few kids started playing jacks. We couldn't believe how much they were getting into it. Eventually somebody brought out a deck of cards. Another dad introduced us to dice. When Larry rolled he'd always say, "Baby needs new shoes!" Things improved at that point. We could get through this, we thought.

TO DUST YOU SHALL RETURN

"The Bible turns out to be a powerful ecological handbook on how to live rightly on earth" (25), says Calvin DeWitt, professor emeritus of environmental studies at the University of Wisconsin. From another perspective, it reads more like a casebook of environmental psychopathology. "So the LORD said, I will blot out from the earth the human beings I have created—people together with animal and creeping things and birds of the air, for I am sorry that I have made them" (Genesis 6:5). "All creatures that swarm upon the earth are detestable" (Leviticus 11:41). "The earth shall be utterly laid waste and utterly despoiled; for the LORD has spoken this word" (Isaiah 24:3). "On that day, says the Lord GOD, I will make the sun go down at noon, and darken the earth in broad daylight" (Amos 8:9). "I will utterly sweep away everything from the face of the earth, says the LORD. I will sweep away humans and animals, I will sweep away the birds of the air and the fish of the sea. I will make the wicked stumble. I will cut off humanity from the face of the earth, says the LORD" (Zephaniah 1:2–3). "For I will no longer have pity on the inhabitants of the earth, says the LORD. I will cause them, every one, to fall each into the hand of a neighbor, and each into the hand of the king, and they shall devastate the earth, and I will deliver no one from their hand" (Zechariah 11:4). "I will punish you according to the fruit of your doings, says the LORD; I will kindle a fire in its forest, and it shall devour all that is around it" (Jeremiah 21:14). "For thus says that Lord: The whole land shall be a desolation; yet I will not make a full end. Because of this the earth shall mourn, and the heavens above grow black; for I have spoken, I have purposed; I have not relented not will I turn back" (Jeremiah 4:27–28). "By my rebuke I dry up the sea, I make the rivers a desert; their fish stink for lack of water, and die of thirst" (Isaiah 50:2). "I will stretch out my hand against them, and make the land desolate and waste, throughout all their settlements, from the wilderness to Riblah. Then they shall know that I am the Lord" (Ezekiel 6:11). "I will dry up the channels, and I will sell the land into the hand of the evildoer; I will bring desolation upon the land and everything in it by the hands of the foreigners; I the Lord have spoken" (Ezekiel 30:12). "I will destroy all its livestock from beside abundant waters; and no human foot shall trouble them any more, nor shall the hoofs of cattle trouble them. Then I will make their waters clear, and cause there streams to run like oil, says the Lord God. When I make the land of Egypt desolate and when the land is stripped of all that fills it, when I strike down all who live in it, then they shall know that I am the Lord" (Ezekiel 32:13–15). "To you, O Lord, I cry. For fire has devoured the pastures of the wilderness and flames have burned all the trees of the field. Even the wild animals cry to you because the watercourses are dried up, and fire has devoured the pasture do the wilderness" (Joel 1:19–20). "I struck you with blight and mildew; I laid waste your garden and your vineyards; the locust devoured your fig trees and your olive trees; yet you did not return to me, says the Lord" (Amos 4:9). "For lo, the Lord is coming out of his place, and will come down and tread upon the high places of the earth. Then the mountains will melt under him and the valleys will burst open, like wax near the fire, like waters poured down a steep place" (Micah 1:3–4). "Was your wrath against the rivers, O LORD? Or your anger against the rivers, or your rage against the seas, when you drove your horses, your chariots to victory?" (Habakkuk 3:8). "Open your doors, O Lebanon, so that fire may devour your cedars! Wail, O cypress, for the cedar has fallen, for the glorious trees are ruined! Wail, oak of Bashan, for the thick forest has been felled! Listen, the wail of the shepherds, for their glory is despoiled! Listen, the roar of the lions, for the thickets of the Jordan are destroyed!" (Zechariah 11:1–3). "Why says the LORD of the hosts? Because my house lies in ruins, while all of you hurry off to your own houses. Therefore the heavens above you have withheld the dew, and the earth has withheld its produce. And I have called for a drought on the land and the hills, on the grain, the new wine, the oil, on what the soil produces, on human beings and animals, and on all their labors" (Haggai 1:7). "Heaven is my throne, and earth is my footstool" (Acts 7:49).

WE EAT WHAT WE KILL

"Abundance and scarcity," said Bob, "have always contended to see which can best destroy human civilizations. In the latter days of our recent empire the two seemed to have joined forces, as the well-off work to wall-off the last bits of the commons. They'll grab up water, airwaves, genes, anything not nailed down. Sorry, check that: they'll pull out the nails and take them, too. A case in point came across my desk the other day: an application from a developer to raze the old meat packing plant and replace it with private lawns and vegetable gardens. I'm all for growing your own food, but these plots won't be like allotments along a railway embankment. We're talking high-cost boutique patches for the urban elites in nearby luxury condos. There're even to be onsite staff gardeners, presumably so that the gentlemen farmer won't have to set down his mint julep and come off his terrace. Dirt under the fingernails? Come on! Now, some of you think replacing that nasty sausage operation is a positive exchange, regardless of how many BMWs are now going to be parking downtown with trunks full of arugula. You've got fresh turnips coming out instead of frantic beeves going in. Upton Sinclair is in heaven grinning. But here's the thing: some of my high school mates were lifers on the wiener line and the bacon table. They were experts in rolling summer sausage and liverwurst. Even olive loaf had its maestros. Where are they now? They lost those good factory jobs, so they're over at Burger King pulling pork for minimum wage. Look at it this way: The wealthy are seizing precious urban land and cordoning it off not just for their exclusive high-rises but now for their personal agriculture! The rich will eat the world, and the rest of us won't even get cold cuts."

THE HUMAN ZOO

The kids were in and out of the Kiwanis pool all morning. The massive park lay in a hidden bend of the river, you'd never know you were in a city, the usual assortment of playground, ball diamond, soccer pitch, and snack bar. The pool itself was well over an acre, lined with concrete, kinda like a giant farm pond with chlorine. My boy was with his buddies out near the middle in the deepest part, which ran only to about four maybe four and half feet. Not much work for the lifeguards today. Occasional running on the deck, ball tossed too hard in face, illicit squirt gun. The minders were looking the other way, mostly. Helluva time. I was suddenly pierced by a memory from many decades before, when my own school had brought us here for a field day. A ghostly echo of a childhood feeling: I'm gonna say "elation mixed with disgust." The pool had been fantastic, but there was gunk in it. Leaves, candy wrappers, bugs, worms, sticks, no doubt freshets of urine. Now here I was, sitting in a lawn chair under a shady beech tree, drinking a grape soda and contemplating life. Time flies. I was supposed to be a parent chaperone, but there wasn't much to oversee. Kids figure out what they need, and that doesn't include hanging with adults. But there was one boy who sat near me at a picnic table, looking at his feet. The poor lad had laid open his big toe. I'd found a Band-Aid for it, but he didn't feel like roaming anymore. He was running out the clock, waiting for the buses to return and tote everybody back to school. He was round as a beach ball, chubby breasts and rolls of flab spilling over his swimsuit, the classic fat kid. I'd watched to see how the others treated him, expecting the usual shunning. But no: the pack had changed since my day. They were nice little humans; they accepted difference, when and if they even noticed it. Still, there was something off about the dynamic; you couldn't rub this many children together and not produce sparks. The spirit of malevolence had to be lurking around here somewhere. But I couldn't figure out the game. Then an odd thing happened, and I found what I'd been missing. A girl had spotted a wounded seagull wandering around the edge of the trees, and a gaggle of kids formed a circle and started taunting it. A few of 'em held sticks and clods of dirt. There wasn't a teacher in sight, so I jogged over and told them nicely to back away from the animal. They looked up at me with their big bright eyes, like the children in *Village of the Damned*. One boy said, "That's a shit hawk. They're a blight on the face of the earth."

MIND AT THE END OF ITS TETHER
(WITH APOLOGIES TO H. G. WELLS)

The end of everything we call life is close at hand and cannot be evaded. Our normal life is happily one of personal ambitions, affections, generosities—sufficiently vivid to outshine any sustained intellectual persuasion of accumulating specific disaster. It requires an immense and concentrated effort to perceive that the cosmic movement of events is increasingly adverse. A frightful queerness has come into life. There is no way out or around or through the impasse. It is the end. Events now follow one another in an entirely untrustworthy sequence. To a watcher in some remote entirely alien cosmos it might well seem that extinction is coming to man like a brutal thunderclap. Our world of self-delusion will admit none of that. It will perish amid its evasions and fatuities. Mind near exhaustion still makes its final futile movement. Our doomed formicary is helpless; the planet spins, climate changes, so that the old overgrown Lord of Creation is no longer in harmony with his surroundings. Go he must.

THE COMPLICATED FUTILITY OF IGNORANCE

Eric and Don Jr. are avid big-game hunters. Leopards, cape buffalo, lions, elephants, crocodiles: the brothers have bagged them all. Dad says, "Eric is a hunter and I would say he puts it on a par with golf, if not ahead of golf." A fourteen-day package safari at African Sky Hunting will run you 39K; for that you receive trophy tags for one lion, one cape buffalo, one gemsbok, one black wildebeest, one springbok, and one zebra. A ten-day elephant package is 35K. The elephant "is a destructive feeder . . . and, rather than adapting to his habitat, he adapts his habitat to suit his purpose." In other words, he is nature's human being. "Hunt this 'big fellow' with the largest rifle you can shoot well." The Weatherby Mark V .460 is a fine choice. The Great Elephant Census conducted by the Paul Allen Foundation showed the species declined by nearly a third between 2007 and 2014. So kill one now or don't complain when they're gone! Eric and Don donate meat from their hunts to local villages.

THE WORLD BREAKS EVERYONE

For over thirty years I've taken my cue from the eminently sober annual Worldwatch Institute reports on the state of the world. The grammatical mood of these essays is the subjunctive combined with optative, which results in a tone of insistence and wishful thinking, as in "Making the Green Economy Work for Everyone" or "Saving the Coral Reefs." One volume, titled *Is Sustainability Still Possible?*, concludes with an essay by science fiction writer Kim Stanley Robinson. In his novels Robinson has often explored possible utopias buried in the seeds of the present. "Is It Too Late?" is a blunt assessment out of sync moodwise with the hundreds of other Worldwatch articles of the last three decades. Robinson arrives at a qualified *no, not too late*—we'll muddle through with a mix of green technology, social change, and Buddhist economics—but the equivocal tone of the essay is remarkably unsettling for a series that has always managed to find silver linings in the darkest clouds. *Is it too late?* Fortunately, whenever a question of this sort is posed—*Should we give up? Is it impossible? Are we crazy to try?*—the answer is invariably in the negative: *No, we mustn't give up*; *No, 'taint impossible*; *No, we're eminently sane*. That's because this type of question is rarely framed in the realis mood (that is, as an actual request for actual facts). Rather, the purpose of such interrogatives is invariably reassurance. They all hark back to the childhood discovery of death: *Am I going to die, mommy? Are you going to die?* The correct response is, *No, honey, you aren't going to die, and Mommy isn't going to die either. We'll be here for a very, very long time, so you needn't give it another thought.* Following this pattern, we might ask: Is it the project of ecological rationality itself that convinces us it is not too late, or is it rather that we are already boxed into the *no, it's not too late* paradigm by the psychodynamics of "hope" and its grammatical servants? After all, death awaits us all, perhaps even tonight in our beds, but we all proceed as if we're immortal. The mind will have its hope. But what is hope, anyway? Hope is magical thinking. It is life lived in the subjunctive/optative mood, the stance of *wouldn't it be nice if it was so?* combined with its faithful implication, *therefore it will be so*. Well, here's to hope! May you live long and prosper! May all your blues be Labatt's! May god have mercy on your soul!

I ALONE CAN FIX IT

At long last the vast left-wing conspiracy was dragged out into the open. Folks could take a gander at its twisted entrails. The moronic inferno burned white-hot and needed to be stoked with fresh victims. Union stooges, climate scientist cabals, radical feminist dike covens, multiethnic cadres of Amurikkka-haters: their time was over. For those, like the new president and the governing party, who found it easier to conceptualize invisible sky-gods than invisible greenhouse gases, there was a moment of Tom Painesque wonderment: They had it in their power to begin the world over again. They licked their chops, the drool running down plump dewlaps a sight to behold. What to do with this broad mandate to drain the swamp? Why not drain some actual swamps? Does the Mississippi require additional dredging? Are the briny, python-infested Everglades still an obstacle to sugar production? What about the interior of the country, its coal and gas reserves straining at the seams with pent-up desire to get out and be combusted? Yet you couldn't fire an AR-15 in any direction and not hit a damned regulation. What wanted restoring was the natural right to emit freely.

Tax cuts would be big league. Tort reform factored in. Industry awaited its unshackling. Government begged for strangling. Czarina Ivanka would advise. In any case, the final proof of our ecological unreason was in the bloody pudding that voters had hawked up and spat toward Washington.

AN ANCIENT POND, A FROG JUMPS OUT, RESOUNDING SILENCE

Professor Zwecklos adjusted his glasses and gazed out over the young men and women in his lecture hall. It occurred to him a summation was in order. He tamped down the tobacco and lit his pipe, to audible gasps. But this was, after all, his last lecture. Retirement was next on the agenda. The smoke-free policy would be suspended for today. "During the term, some of you wondered aloud if our efforts in mitigation and adaptation will come to naught thanks to the flawed nature of our species. I have always moved on from that topic because our business in this room cannot be inference and speculation. As I reflect on a fifty-year career of teaching and research, there have been many similar moments when I've held fast to the proposition that the most pressing questions of all, those that bear directly on our innate yearnings for love, reassurance, and comfort, must nevertheless remain unasked in deference to objective facts. My contention then as now is that good science requires protection from emotion; questions of the human prospect depend so much on social factors about which we are ignorant that it is safer and saner to keep to our knitting. 'I do not respond to hypotheticals,' as the generals and senators like to put it. Still, there are compelling reasons—again, entirely understandable ones—why you may wish me to comment on these matters. And I am not unsympathetic. Perhaps today, a special day for me, I will violate my own rule—just as I am violating our

institution's rules regarding healthy learning environments."
He winked and playfully gestured with the meerschaum. Most
students smiled or laughed; a few wrinkled their noses. "So I
will say this: beyond carbon sequestration, biochar entombment, solar radiation management, albedo modification, stratospheric particle injection, ocean fertilization, and so forth . . .
beyond all these instruments of majestic reason we find, as
ever, the brick wall of stupidity, greed, and arrogance. A powerful triumvirate. An insurmountable triumvirate, in my estimation. Insurmountable: by definition a wall we cannot climb
over. So that is where I land on this issue of the future and its
viability. Yet I confess when I look at your fresh, unlined faces,
I find within me stirring the ashes of hope. Perhaps the latter is biologically inextinguishable. That would not surprise me.
Nevertheless, to reflect usefully on the human prospect I must
separate it from my own prospects, or, for that matter, yours.
Let me tell you a story, or rather paraphrase a story I read many
years ago as a boy. A short science fiction story, to be precise. I
don't remember the author; I don't remember the title. What I
remember is that it involves a couple fleeing before an unnamed
terror. They find themselves in a house, surrounded by that terror. As the story unfolds, it seems that the terror is, of all things,
insects. Insects are attacking people everywhere, effecting by
means difficult to imagine the eradication of the human animal from this planet. As the couple evaluate their situation, they
become aware of a voice. The voice is speaking to them telepathically. It is consoling them; it is telling them that it, too, is an
enemy of the insects. The voice introduces itself. It is the voice
of a lone spider in its web in a corner of the house; this spider is
communicating to them on behalf of the entire class of arachnids. The spiders, it seems, have all along been working to overcome the insects! It is a titanic struggle, protracted and bitter,

fought over vast territories and scales of time, but not without the potential for victory. The spider reveals to them that, with luck and perseverance, the insects can be defeated. Instantly the couple's spirits recover. There is a chance this nightmare will end! The spider announces, 'Have faith. We believe we can save you.' The couple says that this is wonderful news. But, they note, the house is still surrounded. The insects are on the verge of breaking through. Time is short. What is our next move? The spider says, 'Oh, I'm sorry. You've misunderstood. I meant we may be able to save your species. As for you two, I'm afraid . . .'" The professor drew long on his pipe, then began to produce from the bewhiskered O of his mouth perfectly formed smoke rings that floated gently over the heads of his rapt students. They watched as the rings gradually dissipated in the yellow afternoon light admitted by the classroom's streaked and smudged windows. The professor said no more, only continued to puff, and after a number of minutes had passed, the students quietly left their seats and filed out the door.

WE'LL ALWAYS HAVE PARIS

The Conference of the Parties, *recalling, also recalling, further recalling, welcoming, recognizing, also recognizing, acknowledging, also acknowledging, emphasizing, also emphasizing, stressing, recognizing, emphasizing, acknowledging, agreeing, decides, requests, invites, also invites, recognizes, notes, decides, also decides, further decides, requests, decides, welcomes, reiterates, takes note, notes, also notes* with concern that the estimated aggregate greenhouse gas emission levels in 2025 and 2030 resulting from the intended nationally determined contributions do not fall within least-cost 2°C scenarios but rather lead to a projected level of 55 gigatons in 2030, and *also notes* that much greater emission reduction efforts will be required than those associated with the intended nationally determined contributions in order to hold the increase in the global average temperature to below 2°C above preindustrial levels by reducing emissions to 40 gigatons or to 1.5°C above preindustrial levels, *in this context, requests decides, invites, invites, urges, requests, decides, requests, agrees, requests, also requests, further requests, requests, decides, also decides, further decides, invites, requests, recommends, requests, also requests, further requests, requests, also requests, invites, requests, also requests, further requests, decides, requests, also requests, further requests, agrees, decides, also decides, recognizes, decides, also decides, requests, decides, recognizes, invites, recommends, decides, also decides, urges, takes note, decides, requests, decides, also decides, requests, decides, also decides, further decides, decides, requests, also requests, invites, requests, decides, requests, also requests, call upon, invites, decides, also decides, urges and requests, decides, requests, decides, also decides, requests, also requests, further requests, requests, also requests, further requests, decides, also decides, requests, also requests, further requests, decides, requests, also requests, resolves, encourages, urges, recognizes, resolves, encourages, requests, decides, also decides, resolves, decides, acknowledges, welcomes, encourages, also encourages, agrees, decides, also decides, invites, decides, also decides, further decides, decides, also decides, requests, decides, also decides, invites, welcomes, invites, recognizes, also recognizes, takes note, emphasizes, urges* Parties to make voluntary contributions for the timely implementation of this decision.

IN THE FOREST, SOME TREES

The boys and I clambered down the steep hillside to the stand of remnant hemlock. When the area had been logged during the Big Cut days of the late 1800s, hemlock was passed over because, well, it wasn't white pine. Too, this remote little ravine was inaccessible to the sleds the loggers used to pull timber out during the winter months. The boys ran to the base of one specimen and looked up, way up. The tree must have been pushing 140 feet; the branches didn't even begin until 50. *Wow*, said my little guy, throwing his small arms around a portion of its five-foot bole, *This tree must be as big as whale.* The elder one said, *Actually, in terms of biomass, it probably is.* I knew he was right. In a previous summer, I'd estimated this particular tree at 30 to 40 tons, which was about the size of a small right whale. It wasn't much compared to a giant redwood, but then what is? We took the requisite photos: primates leaning on a pillar of carbon. The corrugated bark looked reddish in the dappled light. The air beneath the canopy was warm and dry, with a faint lemony tang. The ground was covered in a bed of soft, flat needles. We felt sleepy, and if it weren't for the mosquitoes we might have lain down for a nap. The younger boy said in a hushed voice, *He must be very old, this tree. I think maybe ten thousand years old. I think he was here with the mammoths and sabre-tooths. A giant eagle had a nest up there at the top, and it looked all the way to the lake and said to the tree, "Thanks for being my throne."* That seemed a good note to leave on, so we slowly made our way out of the grove and continued with the hike. Later I thought about the hemlock throne. There was poetry in my boy's words, and I was proud of that. But he was looking at the trees so differently than I. To him they were life-giving entities of ancient power and permanence. To me the stand was a tiny section of a vast graveyard, and the trees formed their own headstones. We might have been mourners had not our own plots already been laid out in an obscure corner of that grim acreage.

Photo by author

The Narrow Corduroy Road to the Interior

> Dick was finishing up the four-year course in three years and had to work hard, but nothing in the courses seemed to mean anything to him anymore. He managed to find time to polish up a group of sonnets called *Morituri Te Salutant* that he sent to a prize competition run by the *Literary Digest*. It won the prize, but the editors wrote back they would prefer a note of hope in the last sonnet. Dick put in the note of hope and sent the hundred dollars to Mother to go to Atlantic City with.
> —John Dos Passos, *Nineteen Nineteen*

> The ancestral gallery of profound and dogged critics of modernity is long and includes many hallowed names. The best thinkers in Europe have taken this side, including during the twentieth century. In these overpowering analyses one can read how the authors are themselves spellbound by the process they describe. Sometimes a hopeful little chapter is tacked on at the end which is a like a deep sigh at the decline of the world in the face of the general hopelessness, and then the author makes his exit leaving his shattered readers stranded in the veil of tears portrayed. Hopelessness is ennobling, to be sure, and affords the considerable advantage of wallowing in superiority while being relieved of all responsibility for action.
> —Ulrich Beck, *World at Risk*

> All minds seem enslaved by a municipal vocation.
> —E. M. Cioran, *A Short History of Decay*

I am sitting at my desk perusing yet another book about the end of nature when I notice a flutter of wings at the feeder that hangs from an old clothesline stanchion outside my window. I wonder idly how the blue jay I see there matches up to the blue jay in a poem quoted in the book. I put aside the book and examine the animal: blue, white, gray, black, jaunty crest, black collar, long laddered tail, seems to be hungry but a bit sloppy, flinging seeds around with his chunky nib. Lots more detail here, with this by god jay, than when he was just a noun and some adjectives on the page. I decide the poem-jay and the backyard-jay don't sync up. Both place "jay" in my brain, and both give me pleasure, maybe by stimulating the same dopamine pathways. But something lies between them like a darkened strait. I ask myself, not for the first time, what troubles me? Is it that the actual-jay makes the poem-jay a fraud? A plenitude that instantly drains literary description of its value? Or does the poem-jay instead cheapen the flesh-and-blood one? Does our capacity to endlessly assemble words and poems render both jays equally insolvent?

> *snick goes the gray-blue*
> *black oil seed cracks in beak*
> *where is my red pen?*

I know all this musing about real and imagined jays is just a deferral of my other questions, the ones that are less intellectual, more affective, the ones that can bring me to tears. I can discourse on mimesis and representation like any humanities scholar until I'm blue in the face. But that's just a diversion now, a problem rendered lightweight by the gravity of our collective predicament. I know full well that nature is all muddled up with culture, so to speak, and the lines between them, conceptually and physically, are blurred. The questions that I can't answer, can't begin to answer, hurt me to think about. Can what we formerly thought of as nature survive our continual wounding of it? In this age, as it sinks in that the whole planet has become a giant lab experiment gone awry, is any given blue jay more than just another data point? Has our overwriting of nature left us with the capacity to preserve what lies beneath?

on the feeder still
now and maybe evermore
a diminished thing

I decide to exit my study and study these matters more closely. Sometimes the armchair naturalist must go out into the field. Matsuo Bashō wrote, "The gods seemed to have possessed my soul, and turned it inside out, and roadside images seemed to invite me from every corner, so that it was impossible for me to stay idle at home." Now, I'm not a philosopher or a poet, and the gods can't possess a soul I don't have. My answers to the above questions won't satisfy anyone demanding wisdom or beauty or conviction. I can set down words about what I see and think and feel, but they won't add up to much, my brain being so scattered and my sensibilities so dulled. People who spend too many waking hours in books have trouble capturing experience properly. Well, that's the occupational hazard of the English professor. There's a lot of mixing and pouring and general commotion in the kitchen, but when the oven opens and the soufflé comes out, it's usually a fallen mess still raw in the middle.

What I'd really like to do is grok the nature-culture of this town in the time of climate change and environmental decline. Tall order, certain to fail. But effort counts for something, right? I think I might be suffering from what the psychologist Glenn

Albrecht dubs solastalgia: the desolation and pain you feel when you can no longer find comfort and relief in your own place. "Solastalgia is a form of homesickness one experiences when one is still at 'home.'" I remember a filmstrip they used to show us in grade school. It depicted a time-jumping, musical canoe trip around the Great Lakes region featuring a lumberjack-ish fellow who kept finding himself atop a glacier or hanging over a cliff. The yo-yoing geological changes generated by various ice sheet advances and retreats made for comical moments. The film was directed by Bill Mason, himself a famous canoeist and outdoorsman. His contention was that the rising and falling of the Great Lakes as a powerful long-cycle natural phenomenon is now rivaled by short-term accelerated changes prompted by industrial civilization. No argument here. The money-shot is our hero taking a big scoop of lake water in his tin cup, putting it to his lips and drinking deeply, only to spit it out as he realizes belatedly it's full of nasty brown foam. The camera slowly pans back from the lonely paddler, who has found himself floating in a vast patch of frothy chemical pollution. The narrator sings,

> *As you travel on the water*
> *Which the ancient ice age planned,*
> *Just remember there are changes*
> *Which are also made by man.*
> *And the water shimmers golden*
> *In the ever-golden sun.*
> *Yes, I hope there's so much beauty*
> *In the changes that will come.*
> *—Written by Bruce MacKay*

There's something off-kilter and scary about *Homo sapiens* taking on the role of planetary manager, especially since he didn't formally apply and anyway doesn't understand the job description. The human species is a red-faced toddler squatting in front of a computer, holding a hammer and wearing a little apron that says, "Genius at Work." As fixers, we leave everything to be desired.

Well, I guess my point is that we're making it very hard to feel at home in the world. Our dumb necropolis has spread worldwide; its borders in this "shoot-yourself-in-the-head" phase of civilization, the Trumpocene, have met up and entombed the planet's last remnants of wild vitality. So we must investigate the interstices, the popped seams, the little rips in the fabric of our collective winding sheet, to find some solace among all this crummy solastalgia.

I grab a faded cap that says "Cheboygan Chiefs" and head for the garage. Rousseau said his mind worked with his legs; he had to walk to get his thoughts moving. But my decidedly post-Romantic plan is to reseat myself in a car, then zip around looking for clues, clues I hope will manifest themselves in the wildish bits of my hometown, Waterloo, Ontario. Motion is the key: these days ideas emerge for me only at the speed limit. In tranquility, I just nod off. I repair to a nondescript four-door Civic. A 1.5-liter fuel injected motor. I'm aware that the entire automotive world is noncarbureted now, so "fuel injection" is an unnecessary qualifier, like saying "this car comes equipped with wheels." Still, fuel injection sounds efficient and indeed this heap gets thirty-five miles to the gallon, though it burns a bit of oil.

a sip among gulps
jet black Honda '97
hey I do my part

As I back out, I see in the mirror the gaunt form of my neighbor waving at me from his driveway across the street. I roll down the window as I wheel into the street.

"Hi Phil." I notice he is barefoot and wielding a rusty wrench.

"Morning. Did you hear about the ornithologist who decided to study the Carolina parakeet? When informed it was extinct, he replied, 'Really? How auk-ward!'"

Phil is a retired physics professor and active punster. "Beautiful!" I say, and mean it. He knows my tastes. I roar off (the Civic's muffler is developing a modest hole). I turn the radio on to cover the rumble. Talking Heads, "Life During Wartime." I roll through the stop sign at the end of Dunbar Road, then ease out onto Erb Street, which will take me right to my first stop: Waterloo Park.

But first I have to get through the old Seagram's Distillery grounds (and a long digression on science, local history, and modern communication). Once part of Waterloo's beer and booze infrastructure, this area of downtown contains an international governance think tank, a bunch of renovated lofts, a flurry of fancy shops, and a murder of mediocre eateries. Across the street, on the site where once stood a barrel-yards and the Waterloo Memorial Arena, there is a long black shoebox called the Perimeter Institute. Inside, mathematicians and physicists are writing equations on chalkboards that may have something to do with the nature of the universe, or perhaps the structure of quantum computers. I don't follow that stuff anymore, though I once thought books like *The Tao of Physics* and *The Dancing Wu Li Masters* held the keys to all life's mysteries. Or maybe compounded them. Anyway, a couple of wispy memories of my early minor hockey triumphs play across my synapses, foiling any attempt to imagine what a grand unified theory would do for me personally.

the universe is
not what it was: dense pinhead
of possible fire

Next, I have to maneuver around some serious construction for a new light rail transit line that'll run down the middle of our main drag, King Street. (In every Canadian city, you'll find a King and a Queen Street. Also high on the list are Victoria, Wellington, and Maple.) As the street was dug up, a deeply buried hundred-foot-long corduroy road was uncovered. It was inspected by local archaeologists, who dated it to the late 1700s or early 1800s. It was probably contemporaneous with the original settlement established near the mill of Abraham Erb, a Mennonite from Pennsylvania. The Mennonites were a splinter group of Anabaptists. Anabaptists were one of the many sects to emerge during the Protestant Reformation. They opposed the practice of baptizing infants. I guess they still do. Mennonites from Lancaster County arrived in Upper Canada after the American Revolutionary War. Prior to the Mennonite influx, the area was held by the League of Six Nations, the Iroquois confederacy that was gifted the land by the British in exchange for their help against the Americans. Before that, Seneca and Mohawk tribes had controlled the area. And before that it was inhabited by a tribe known as the Attawandaron. They were part of the Neutral Confederacy, so called by the French because it attempted to stay out of the way of contending Hurons and Iroquois. Like the Mennonites, the Attawandaron were a peace-loving people. During the Beaver Wars of the mid-1600s, the Attawandaron were annihilated. At that time, the area was one of the most densely wooded parts of North America. Some of the trees that made up the old wooden road could have been saplings when the last of the Attawandaron expired under the assault of hostiles, famine, and disease. The relic logs were taken to the local landfill, cut into two-foot sections, and distributed free to interested citizens on Saturday morning. They were gone in twenty-seven minutes.

two hundred pieces
of history—folks figured
they'd varnish 'em

Safely skirting the excavations, I proceed around the bend past the shopping center, the former cop shop, the public library, the Habitat for Humanity headquarters, a plastic surgeon's office, and into the old section along Albert Street, named after the prince consort. A nice part of town here: Victorian houses, well-kept yards, and tall maples. I hang a quick left and approach the entrance to our city park. The car is performing well. I roll down the windows. The air is cool—it's mid-May—and at this low speed the busted muffler is civilized enough. There are a few park patrons strolling about and a young couple on a shaded bench thumbing their handhelds.

On that latter note, maybe I should mention that Waterloo is home to RIM, the makers of Blackberry communication devices, which spawned all subsequent smartphones. Yes, you can thank Waterloo. The company helped revive the local economy after the older base of felt manufacture, distilling, brewing, rubber, and auto parts bit the dust. But given RIM's all-but-inevitable destruction in the face of overwhelming competition from Apple and Samsung, Waterloo may one day become the Flint, Michigan, of the information economy. For now, folks are still upbeat.

Every man, woman, and child has these gadgets around here, or so it seems. So far, I've resisted. Don't own one. Refuse to comply. They are another nail in the coffin of the world and our ability to function as plain members within it. Aldo Leopold never had one and got by just fine. When I go to the rink, half the people in the stands are tapping on their phones while their kids play. A minority are still yakking into them. I sat outside a Dairy Queen a few days ago and listened to a conversation a mother was having with her ten-year-old son about a friend of his whose parents were divorcing and the vagaries of custody arrangements. It was a remarkably dumb and self-serving monologue by the mother, mostly to the effect that everybody should do whatever makes them happy. Sounded like she was plotting to ditch the boy's father, was my guess. Suddenly, she trailed off, dialed a number on her phone, and began talking to some friend or acquaintance. Her child continued to lick his cone while she babbled inconsequentialities to this remote person. Then she switched the phone to loudspeaker, mounted her bicycle, and pedaled off, her son struggling to keep up on his bike, as the static-jammed voice of the phone buddy burbled across the parking lot. The boy had a blank look on his face; he was probably eager to get home to his video console.

I suppose I should be thankful that texting and social media are starting to eliminate the need for such cringe-worthy public conversations. But texting and social media represent an even more powerful threat to co-presence. They draw one's head deeper into the virtual sand. Augmented and mixed reality will seal the deal. When folks can overlay boring everyday life with gold coins, flaming barrels, and nude celebrities, who'll want to come up for air? It'll all seem normal and natural before much longer.

the sound of gunfire
off in the distance I'm get-
ting used to it now
—Written by David Byrne

Anyway, when the old booze infrastructure was shut down, the chatterbox one started up. Contact. People love to talk. As soon as they're let out of class my students get into their bubbles. They're triangulating with other kids also let out of class, trying to get a bead on a meet-up. Sometimes they come together in the Arts Quadrangle, still texting to one another on the phone as they meet f2f, their bubbles merging at last. It reminds me of the old gag in *Get Smart*, where the chief and Max talk under the Cones of Silence to protect against eavesdropping. But these smart-cones create silence in a different way. They shut off verbal contact by rendering that medium uninteresting. Nothing so sad as the restaurants near campus, many now quiet as mausoleums, where chums gather for lunch and then spend their time glued to their smartphones. Or dumbphones, as they should be called, not just for the obvious reason. Old-fashioned conversation can't compete for engagement and excitement with screen-based media. I understand that when one hears the ping of a new message, a little squirt of dopamine washes across the brain's pleasure center.

> *glass breaks cymbals crash*
> *tones of Timberlake or Swift*
> *the phone tolls for thee*

But I'm not on my nonexistent cell phone. I'm supposed to be thinking about the nature-culture dialectic or some such. The inner curmudgeon must go sit in the corner. I am driving into Waterloo Park now, this gently rolling, gently forested little jewel in the center of the city: I need to focus. One hundred acres, all told, with a dammed-up lake and a little stream, the former called Silver and the latter Laurel. Laurel leaves the lake at the east end, plummets down a fake waterfall and into a concrete tunnel that runs a few hundred yards under a perpetually redeveloping shopping mall. There are some nice dirt walking trails, trees in abundance, plenty of grass, even a cricket pitch in addition to the usual ball and soccer fields. A splash pad has replaced the outdoor pool I remember running around back in the day. The magnet part for kids is still the zoo. It is not now and never has been much of a zoo: just a few rabbits, some peacocks, a fenced-in paddock with miniature donkeys, llamas, a fallow deer or two. But on any day of the week you'll see beaucoup tots wandering around, peering through the chain link, feeding bread to animals under signs that say "Do not feed the animals," accompanied by dutiful parents and caregivers who make the usual noises of encouragement and condescension (when they chance to look up from their smartphones).

I take the single-lane drive that skirts the east side of the park, turn left toward the Waterloo Tennis Club that pokes in from the north side, then up a rise—cricket pitch to the left, clay courts to the right—over the brow, then slow near the bottom next to the baseball diamond, the very one where I pitched my first little league game. Just off my port fender is the far end of the line of zoo enclosures. I don't even have to exit the car to engage memory. The closest cage is the most secure, with ground to ceiling chain link and concrete bunkers. It's empty now. There's a faded nameplate screwed to the fence that references a certain pig that's obviously gone on to its reward, perhaps to Piller's, the local cold cut plant or, as I hope, a cozy mud yard at a local farm. I'm no expert on the pig era of the zoo—that was a relatively recent thing. What I recall, from back in my salad days of minor baseball and hockey, were the black bears. My mother took me to visit them often. These poor devils lolled around in their cramped quarters, bored stiff, like most zoo animals. A log and a bale of hay constituted their furnishings. Mostly they slept. I remember Mom tsk-tsking, saying, "Poor things." Now, lots of folks were getting the anticruelty religion at that time. There was a steady hue and cry in the letters to the editor section of the local rag. Finally, must have been during the first Trudeau administration, the bears were sent packing back to the wild or, more likely, to another zoo. Bears in captivity don't flourish—like most living things in captivity besides maybe fish. But they also don't flourish when released. They usually have no idea what to do. You wouldn't know that from *Free Willy*.

> *empty concrete cell*
> *branches rattle bars in breeze*
> *can not feed the bears*

I read of a village in India of 6,000 that shares its space with 3,000 cobras, which the villagers also revere. The snakes crawl in and out of the houses, under chairs and into beds, over arms and between legs. We've got maybe 150,000 humans in this town, and we don't worship anything along the animal line. Unless it's dogs and cats. They're all over the park today, on their stretchy leashes, dragging their owners, or what I call shit-baggers. It's a funny thing, this new domestic arrangement of the last 10,000 years. From what I understand, dogs are the only animals that actually like humans.

So the controversial bears are long gone, any desultory hankerings for wildness they may have aroused in Waterloovians faded from our emotional universe. Even in those days, folks checked out the bears, then went for sno-cones at the park concession. I doubt there were many epiphanies. I check my watch, then throw my car into gear and retreat from the park through the back entrance. It's a short hop over to my university where I ply my trade, such as it is. I'm one of those rare academics who, for good or for ill, got a job back in his hometown. The humanities hall is a sprawling brown brick warren, the old-guard disciplines splayed out down its dim passages. Yes, there's a shiny new accounting and finance wing with a sunny atrium, but that seems out of bounds, as if designed for a newer breed of people. I can't resist stopping in the fifteen-minute parking slot and running up for my mail. There isn't any. No news is good news. It's Sunday, by the way, and the place is as lively as a grave. I like the quietude, the solemn, closed office doors and echoey, vacant hallways. Excellence has the day off. I am grateful to be working here, happy with my department and colleagues. I enjoy the last best job in the world, and like most non-STEM professors I hope I'll have shuffled off into retirement before the lit-and-language-gig winds down forever or morphs into the Department of Soft Skills. I meditate briefly on our significance.

dusty yellow air
bookshelf sags against dun wall
old words for old ways

Back at the car, the radio says, "Canada's diplomats are now agents of commerce, according to the Ministry of Foreign Affairs." You mean they weren't already? "In Alberta, a new pipeline proposal pits producers against aboriginals." CBC, our national broadcaster, insists on calling the tar sands the "oil sands," aligning its usage with official government nomenclature for the solid mixture of soil and hydrocarbon currently being strip-mined in the subarctic taiga. That sort of greasy rhetoric is what put Plato in a dither 2,500 years ago. At least they fall short of calling it black gold.

From my building, I zip around the ring road, passing by the fancy new admin ziggurat, the quantum-nano center with its optically disturbing window treatments, and a gaggle of Canada geese grazing on the grass along the boulevard. Then across Columbia Street into what we affectionately call "the north campus." The north campus isn't really a campus. It's what's left of a 600-acre farm that for the longest time remained undeveloped, even as the city came to encompass it on all sides. In the last decade, however, the north campus was rezoned for various town-gown joint ventures, the keystone component being the David Johnson Technology Park (DJTP). Johnson was a successful president of the university, and then went on to become Canada's governor general. The GG is technically the highest-ranking government official, the Crown's viceroy and de facto head of state of our dominion. But his basic job, much like the modern university president, is to travel about giving speeches, handing out awards and medals, and hobnobbing with visiting potentates and other elite constituencies.

When I was younger, even before I took a job at the university, I liked to ride my bicycle around the woods, streams, and fields of the north campus. At an old abandoned barn near the center of the property, I discovered an ancient butternut tree. I took a nut home and tried to get it to sprout. No luck. That butternut was the only one I've ever encountered in these parts. Uncommon around here, unlike its ubiquitous cousin, the black walnut. I like to imagine that the original homesteader brought the seed with him when he wagoned up to Ontario, possibly from the Pennsylvania Colony, and planted it near his farmyard. The butternut wasn't great as lumber, its wood softer and weaker than walnut, but the fruit had many uses: as food and, as that great Puritan Roger Williams noted, for little niceties like buttons and watch fobs. He mentions that the Indians "of these Wallnuts . . . make an excellent oyle good for many uses, but especially for their annoynting of their heads" (quoted in Marlow 103). My butternut was cut down about the time they were breaking ground for the various roads and buildings entailed to the tech park. These days we anoint our heads with other products and tell time with fobless smartphones. Of this tech park, I can rightly claim it is now more Internet than butternut, more software than soft wood.

around here the trees
are made of gold and logic
the new bright forest

Yeah, I'm aware that my preference for the "original" farm over the tech park is based on a specious history. The old farmstead was itself a square mile of devastation, not quite as ruinous as a glacier but in that ballpark. Before it was cleared for crops, what stood here was a primeval, broadleaf woodland, part of the vast Carolinian forest biome that stretched shore to shore across the lower Great Lakes region. In it lived bears and wolves, passenger pigeons and eagles, chestnuts and elms, the Attawandaron and the Mohawk. Gone now, extinct or extirpated all across these fully technologized, commoditized territories.

Was that farm poorer in its occupation of this landscape than the original forest? Yes. Is the new tech park poorer in its occupation of this landscape than the farm? Well, is an acre of corn poorer than an acre of building and parking lot? No. The land might as well be sealed in a tomb when it's under concrete. But what about 600 acres of corn versus 600 acres that in addition to 200 acres of buildings, roads, and parking lots includes a significant environmental reserve, green spaces, and ponds? That's a more complicated footprint calculation. I only know that I miss the butternut.

From the DJTP I blast down Columbia through the former student ghetto—now the student condo canyon, thanks to a recent spate of urban densification—across King Street, past the McDonalds and the tire store, across Weber Street, past the demolished La-Z-Boy factory, through a green strip, over an overpass, out into the town's original leapfrog extraurban development, Colonial Acres. This was where I spent my first eleven years. They were "acres" all right, but don't ask me about the "colonial" part. That's an Americanism. The settlement of Upper Canada didn't have a "Colonial Era," though we were a colony par excellence, one much more obedient than its southern cousin. No revolution here, just a gradual repatriation of powers. Really, I think you need a colonial period that ends decisively and affirmatively to be able to refer to it nostalgically. Our naming traditions normally produce monikers like "Loyalist Parkway," "Commonwealth Plaza," and "Monarch Woods."

But here it is anyway: the inapt Colonial Acres. I turn left and right to reach the short of *S* of Whitmore Drive, site of much activity when it and I were new. Every house then a rookery of energetic kids: the Stabenows (four boys), the Fergusons (six mixed), the Diamonds (five mixed), the Lawsons (three boys), the Carlisles (two boys and a girl), the Bells (boy and girl), the Kadwells (boy and girl), the Parkers (two boys and a girl), the Belaks (three boys), the McMurrys (four boys). Yes, it was Baby-Boom Centrale. The road belonged to the trikes and Big Wheels, and the yards were lousy with sandboxes, swing sets, and jungle gyms. (But, my, how the neighborhood has aged! I have a friend who moved onto the street a few years ago: He noted but a paltry three trick-or-treaters on Halloween.)

In those days, this street was an outpost of Waterloo, separated by a mile or more from the city, a little pocket of ranches, French Provincials, and Tudors surrounded by Mennonite farms. There's our house, the first or second one built, sitting on its acre lot. My folks commissioned it in 1963, not long after they moved here from Tallahassee, toting my three older brothers in the back of a Dodge Dart station wagon. I understand it was a pretty miserable transition for all of them, that shift from Sunbelt to Snowbelt, with much second-guessing and at times even downright despair over the harsher climate. I, an autochthonous Canuck, had no trouble bearing seven months out of twelve under imperfect skies. I spent my formative period in Waterloo, though we shifted back and forth to Michigan each summer, and later I would live many years in Florida, Georgia, and Indiana. Our whole family felt more expatriate American than Canadian, and to this day, as a dual citizen, I'm equally drawn to the politics and cultures of both nations, with the net result that I'm perpetually whipsawed between two dangerous visions of the future of North America. Take it from me, there is no safe haven. The world's longest undefended border sutures two countries that are both going over an ecological cliff, hand in hand like outlaws hoping to escape hanging but really just joined in an unspoken suicide pact.

The mature oaks and walnuts are still standing, noble as ever, in front of the faux-antebellum-style front porch. (Thank god my parents never installed a lawn jockey, though I recall there were plenty of them around the neighborhood.) There's my room, second in from the right on the top floor, actually a paneled den, probably where my dad imagined himself one day sitting in a leather club chair after I, the baby, finally moved to one of the downstairs bedrooms when an older brother went off to college. At that time, encroaching on the left side of our yard just over the split-rail fence were The Woods, an untamed riparian strip, probably not much more than thirty yards from side to side, but as good as the back of beyond when I was six. Thick with cedars, spruce, and willows. My friends and I spent much of our playtime among those trees. They were our purlieu; no adults could enter those deep recesses. The nameless little creek itself was shallow and sandy, teeming with minnows, frogs, and turtles. Its waters must have been relatively clean-running, because I remember a knockdown, drag-out fight with one of my brothers involving fistfuls of watercress, a plant that grows only in the best streams. Dam building was as natural as breathing. One of the neighbors a few yards down had impounded some of the stream into a side pond that he stocked with beautiful rainbow trout, which provided endless fascination and maybe a few after-dark, illicit fishing trips.

rough bark wet willow
soft needles lemon scent dusk
shrieks come to dinner

I slow at the head of the crescent and gaze at my old demesne. The Woods now appear perfectly ruly, trimmed and raked, the understory planted with shade-loving hostas and elephant ears, the recesses exposed and rendered adult friendly. Past the farthest corner of the yard near the ingress to The Woods I can just make out the now-empty space where once stood the kennel of Rex, a neighbor's terrifying German shepherd. I don't recall ever seeing Rex out of that pen; he just padded around in there, leaping at the fence and baying like a fiend whenever we went near him. Daring each other to test the limits of Rex's sensory apparatus was blood sport. Within ten yards he was aces, the Cerberus at the gate of that forest, just one head but the racket of three.

Our own dog, Chipper, must have filled Rex with jealousy, for Chipper roamed freely through Colonial Acres, his days his own to enjoy. My parents were not early adopters of leash laws, having grown up in an era when dogs were trustees, not convicts. Chipper was a local legend, a dog with no boundaries and no sense of his place, who followed his nose wherever it took him, which was most often into other yards and other gardens. You might find him sleeping blocks away in somebody else's garage. Chipper had at least nine lives: lost and found so many times it was a credit to my parents' commitment to his inborn liberty that they never tied him up. I put together only many years later how much loathed he was by neighborhood parents and feared by their children. Chipper, amiable beagle of paradox, who could revert to wild-thing if disturbed while in repose. I recall an image of my pal Drew and me playing in my sandbox, the two of us road building in the beautiful moist sand my father periodically trucked in for me from Lake Huron beaches. Chipper snoring in a corner. In my memory, as fresh as the day it happened, I say, "Don't pet him, he won't like it." Drew ignores me, the dope. And then the snarl, the snap, and Drew's forehead laid open to the bone, the ragged flesh hanging like strips of pastrami. He staggers home screaming. Chipper goes back to sleep; I wander inside for a snack. Even in those days, Chipper could not have survived such a display of child harming, yet so far as I know nothing came of this incident. Do I exaggerate it? Perhaps the vividness of my recollection is testament to the waking daydream that is much of our childhood.

I visit these precincts at least once a year to renew our acquaintance and witness any changes; the pink-bricked house in turn looks at me and notes the growing paunch and receding hairline. I never see the current owners. They look to be perpetually on vacation, for the curtains are always drawn and there is no activity in the yard. Come to think of it, a lot of houses look uninhabited these days. We spend so much of our time in kitchens and media rooms that life seldom spills outside anymore. Canadians are an indoor people anyway. Winter runs at least half the year, and when summer arrives we need time to train up and make good use of it. By September we're as tanned as Australians—but then October and November come along and drive us back into our wired hibernacula.

Now it's time to move on to my fluvial haunts. The Grand River drains a large portion of southwestern Ontario and is the biggest and most famous of our modest waterways. Its source is to the north in the highlands near Georgian Bay, but it flows south past Grand Valley, Fergus, Waterloo, Kitchener, Galt, Paris, Brantford, and Caledonia before entering Lake Erie at Port Maitland. Steamships once plied its waters up to Brantford, but by Waterloo it's unnavigable by any vessel with more draft than a canoe. Strangely enough, our town was built about three miles from its banks, up a narrow tributary that apparently made for a better millworks. So for many years the Grand was barely a feature of the town. But for those of us in Colonial Acres, it was but a half-mile jaunt down a dusty road to arrive at its inviting waters, which slip noiselessly alongside the primitive park where I'm now idling: Kaufman Flats. Although the approach road has been paved, the tractor-tire swing gone, and the little troll bridge replaced by something sturdier and liability proof, it's still a place of rich memory, still enclosed by the same curtain wall of maple, willow, and walnut mounting the high bluffs. The territories on the other side—wetlands and distant farms—appear as unreachable as ever. Rivers are like that. They resist our advances, throwing out new loops, drowning fields, carrying away earthworks and flimsy structures along their banks, renewing themselves flood by flood. They resist the identities we try to impose on them. Merwin writes, "the river still seems not to move / as though it were the same river (*Sirius* 96). Sure, you can straitjacket and dam any river you like, but water seeks its level. So the Grand stays kinda wild despite our efforts to master it.

Not much of a Mississippi, I know, but with its own riverine mojo, especially for a kid off the leash. Fishing for suckers. Gigging for frogs. Swimming in its quiet pools. Tobogganing down its bluffs. Floating through the broken dam. Skipping stones along the flats. For all my life, I've spent my summers on Lake Michigan; the stretch of the Grand I know is an elongated puddle by comparison. But I'm a fan of the river nevertheless. I like its unflustered flow, the amiable press of pea-green water on the banks, the shoulders of goldenrod and milkweed, the gravelly islands that rise and sink with the seasons. All of us are drawn to rivers and lakes and oceans. Our best home places are those on high ground with a prospect of water.

smooth surface glistens
a muskrat swimming—no wake
the far shore beckons

North and south of Kaufman Flats the overlooks are now lined with fancy townhouses and detached mini-manses. The farms are gone, their orchards and fields plowed under to manure a hundred new Whitmore Drives. The river is no longer marginal. The town, after two hundred years, has finally rediscovered the source of its existence, this blue highway that brought natives and trappers alike into a one-time wilderness. I suppose Kaufman (German for "businessman") Flats are at last fulfilling their destiny on Planet Commerce. Long ago, I attended a talk by David Suzuki, Canada's premier environmentalist and science popularizer. Suzuki spoke of his youth near London, Ontario, back in the 1940s. The land has changed so much since those days, he'd said. Woodlots cut, swamps drained, farmlands developed, roads extended, chemicals applied, species extirpated, nature rolled back as humans roll forward. Kids growing up there have little opportunity to interact with nature, every last wild space and leftover commons eaten up by the growth machine. It's the story we all know. It's our common heritage.

acres of side-splits
parking lot reeks of new tar
trash floats in duck pond

From the flats I head out to the hinterlands along the old river road. It's been turned into an extension of University Avenue, which is now a long way from the university and runs north–south here instead of east–west. I know if I follow it far enough it'll curve back west again and chase its own tail. It's been rerouted to account for all the new subdivisions, as well as the giant recreational grounds that have been laid out at the (former) edge of town. This is our new jewel, called RIM Park—again with the smartphones!—which comprises a links course hard by the river, a miscellany of artificial and natural turf fields, hiking and cycling paths, and a giant indoor facility that contains a soccer pitch, four ice rinks, a couple of gymnasiums, workout rooms, a bar and restaurant, a sports medicine clinic, along with umpteen meeting rooms, banquet halls, and offices for various administrators and sporting organizations. I've spent a lot of time in this building, coaching my boys' hockey teams and playing myself. The ecological footprint of this park is immense: the chillers for the rinks working against the heaters for the rest of the building; the vast playing fields mown and fertilized and weed-treated; the hectares of tarmac plowed and resurfaced; the farms displaced, the wetlands managed, the woods reconfigured, the whole Suzuki litany. I have a love-hate relationship with RIM Park: love what it allows me to do, hate its environmental cost. In other words, I have the same feeling toward it that I have toward our civilization as a whole. Neither have a bright future.

wind through broken glass
vines entwining rusted stands
Ozymandian

I take a circuitous route through the park and yet another brand-new subdivision, and thence onto the old road to Conestogo, a little farm town increasingly a bedroom community for Waterloo. In Ontario, thankfully, zoning is very strict. The city stops on a dime at the city limits. This is some of the best agricultural land in the world, and it must be safeguarded. More or less. I leave behind the last bit of suburbia. I pass by a one-room Mennonite schoolhouse, still in operation, then plunge over the brow of a steep hill past a big red barn. Laid out before me is a compact valley. The river hugging the far bluff is the eponymous Conestogo, which will be joining the Grand about a half mile downstream. I drive across the floodplain, noting the fields to each side, which are brown canvases at this time of year but will soon sport a green fuzz of corn or soybean shoots. The river has been carving out this valley for ten thousand years, dutifully bringing soil and bearing it away, forming the rich bottomland that attracted so many German settlers.

I pull into a little lay-by along the lead-up to a one-lane bridge. Rusty iron-trussed relic from the year Karl Benz introduced the motorcar, no doubt slated for demo in somebody's master plan. To be replaced by a characterless slab of cantilevered concrete. As I carelessly throw open the door and step onto the shoulder, a shiny new Civic thumps off the bridge and whizzes past me. For a split second I see the woman's face, her expression all surprise and irritation. She wonders what business I have here on this back road. Good question. The gravel grinds under my feet, the wind tries to whip off my cap. I'm out of my climate-controlled cockpit, initiating contact with the actual earth. I walk to the middle of the bridge's formerly planked, now rough-paved surface, then peer down at the glassy waters of the Conestogo. It's high at this time of year, still flush with spring rain; it'll lower as the summer wears on. One year it might lower and never again rise. It might become a dry groove in a dry land. Who knows? In the shallows a few yards downstream, a blue heron is stalking fish. I'd like to see it strike; I don't recall if I've ever seen that happen.

I'd charged myself to contemplate real and ideal nature, but I realize, not for the first time, I have nothing useful to contribute on that subject. Just a lot of tangled thoughts that periodically emerge as tangled sentences. Sure, I try to be open to universal currents as much as the next guy. But I'm the worst kind of transparent eyeball: lots of light shooting through and not much falling on the retina. Consciousness is feeling more and more like a few twitching neurons and some irritable mental gestures, to paraphrase Lionel Trilling. Well, I comfort myself that I'm a suitable token of the human type. In the history of thought many distinctions have been made between real and ideal, mind and substance, culture and nature, and so on, for reasons that turned out to be temporary and fungible. Conceptually, we can pull nature and culture apart, mash them together, or wring them of meaning like old dishrags. Whatever. Fill your boots. The human curiosity cabinet is an accretion of mindstuff over time, each generation rearranging its mental knickknacks as it wishes or must.

Always, it's the big ideas that occupy the most space and draw the keenest attention. We so-called intellectuals spend lots of time poring over the writings of an Aristotle, Emerson, or Heidegger in the hopes of descrying an ecological strain that may lurk under the anthropocentric surface. We assume that a reconstruction or recovery of such thought can spur us forward, give us more ammo to argue for a better future based on those titans' timeless wisdom. Lost in these lucubrations is what was never found in the first place: the outside of thought, the unknown unknowns, the endless stupidities of the human animal that have as much to do with who we are than anything we can positively know. This ocean of ignorance explains why, in this most buoyant phase of civilization, we are going nowhere but down.

Maybe what needs to be thought is how useless thought has been. "My head is hands and feet" (*Walden* 96), said Thoreau, a T-shirt-sized apothegm for the embodied nature of mind. In that spirit, I'm advising myself to move on from airiness to terrestrial matters, to things that are graspable at the level of feeling and affect, that don't require authentication by an overwrought brain but rather simply its acquiescence. More pressing, for example, is the question of places, now and in the future. We excel at destroying them, no question about that. And the way we go at the destruction hammer and tong you'd have to think we relish it. At the same time, there may be something in us that wishes to hold fast to the places we've known. Perhaps that's what took hold of me today. If you need a name call it *topophilia*, following the usage of the geographer Ye Fu Tuan. Or call it nothing at all: just feel places tugging at your hands and feet. Let your head follow.

Places are where our memories linger; perhaps places are memory's true anchorages. Places are outside of us, yet equally so inside. But places are slipping away—faster now than ever before. We've put ourselves in charge of their upkeep, but we've yet to do the hard work of figuring out what that actually entails. Maybe we just can't do that sort of work, it's beyond us. Witnesses to our own incorrigibility and incapable of self-correction, we carry on, diminishing all places to versions of Colonial Acres, or the tech park, or the bear cage. And in this way every new place-memory we make already has one foot in the grave.

My own little funeral route through the places of my early life now complete, the only thing I'd like to get a grip on is this river and this heron. The bridge keeps me above them, but at least it gets me near them. I shouldn't stay here long; I've got a thousand things to do. There's never enough time. But I will stand here for a while. I will hold. I will dam the flow of words in my head and attend to what I'm seeing below. I will be part of this scene and no more. Let an observer make of me what she will.

green water slides by
the stick legs of the wader
who waits and watches

look, now, the slender
arc of bill flashes, spears, and
rises with silver

(un)witting witness
is ever empty and full
as bird and river

Bibliography

Adorno, Theodor, and Max Horkheimer. *Dialectic of Enlightenment*. Trans. Edmund Jephcott. Stanford, CA: Stanford University Press, 2002. First published in 1944.

Aliens. Dir. James Cameron. Twentieth Century Fox, 1986.

Albrecht, Glenn. "'Solastalgia,' a New Concept in Human Health and Identity." *Philosophy Activism Nature* 3, no. 41–44 (2005).

Appleton, Victor. *Tom Swift and His Electronic Hydrolung*. New York: Grosset and Dunlap, 1961. Project Gutenberg, October 15, 2017.

———. *Tom Swift and His Giant Telescope*. Racine, WI: Whitman, 1939. Project Gutenberg, October 15, 2017.

Baker, Carolyn, and Guy McPherson. *Extinction Dialogs: How to Live with Death in Mind*, Kindle loc. 1928–33. Tayen Lane Publishing. Kindle.

Bauman, Zygmunt. *Liquid Fear*. Cambridge: Polity Press, 2006.

Beck, Ulrich. *World at Risk*. Cambridge: Polity Press, 2009.

Becker, Ernest. *The Denial of Death*. New York: Free Press Paperbacks, 1997. First published in 1973.

Beckett, Samuel. *The Unnamable*. New York: Grove Press, 1958.

Bierce, Ambrose. *The Devil's Dictionary*. Project Gutenberg, October 15, 2017.

Bitzer, Lloyd. "The Rhetorical Situation." *Philosophy and Rhetoric*, no. 1 (1968): 1–14

Burke, Kenneth. *A Rhetoric of Motives*. Berkeley: University of California Press, 1969.

Bush, George W. "First Inaugural Address." Delivered January 20, 2001. Bartleby.com, October 15, 2017.

Calinescu, Matei. "The End of Man in Twentieth-Century Thought: Reflections on a Philosophical Metaphor." In *Visions of Apocalypse:*

End or Rebirth?, edited by Saul Friedländer, Gerald Holton, and others, 171–95. New York: Holmes and Meier, 1985.

Children of Men. Dir. Alfonso Cuarón. Universal Pictures, 2006.

Cioran, E. M. *A Short History of Decay*. New York: Arcade, 2012. First published in 1949.

Clinton, Bill. "First Inaugural Address." Delivered Wednesday, January 21, 1993. Bartleby.com, October 15, 2017.

Cohen, Tom. Introduction to *Telemorphosis: Theory in the Era of Climate Change*, vol. 1. Ann Arbor, MI: One Humanities Press, 2012.

Commoner, Barry. *The Closing Circle: Nature, Man, and Technology*. New York: Bantam, 1972.

Costa, Rebecca D. *The Watchman's Rattle: A Radical New Theory of Collapse*. Philadelphia: Vanguard Press, 2010.

Deepwater Horizon. Dir. Peter Berg. Di Bonaventura Pictures, 2016.

DeWitt, Calvin. "Reading the Bible through a Green Lens." In *The Green Bible* (New Revised Standard Version), I-25–I-34. New York: HarperCollins, 2008.

Di Giacomo, Frank. "Q&A: *The Road* Director John Hillcoat on Adapting a Modern Classic." *Vanity Fair*. Vanityfair.com, June 15, 2011.

Dos Passos, John. *Nineteen Nineteen*. New York: Signet, 1969. First published in 1932.

Duell, Mark, and John Steven. "The End Is Really Nigh." Mail Online, November 11, 2011.

Dupuy, Jean-Pierre. *The Mark of the Sacred*. Trans. M. B. DeBevoise. Stanford, CA: Stanford University Press, 2013.

Edelman, Lee. *No Future: Queer Theory and the Death Drive*. Durham, NC: Duke University Press, 2004.

Eder, M. D. "The Myth of Progress," *British Journal of Medical Psychology* (1962): 35, 81.

Ehrenfeld, David. *Swimming Lessons*. New York: Oxford University Press, 2002.

Fernández-Armesto, Felipe. *Civilizations*. New York: Free Press, 2001.

Ferris, Timothy. "The World of the Intellectual vs. The World of the Engineer." Wired, October 13, 2011. Accessed October 15, 2017.

Fitzgerald, F. Scott. *The Crack-Up*. Ed. Edmund Wilson. New York: New Direction Books, 1945.

Franzen, Jonathan. *The Corrections*. New York: Farrar, Straus and Giroux, 2001.

Gore, Al. "Nobel Lecture, 10 December 2007." Nobelprize.org, October 15, 2017.

Gray, John. *The Silence of the Animals*. New York: Farrar, Strauss and Giroux, 2013.

Guerrasio, Jason. "A New Humanity." Interview with Alfonso Cuarón. Filmmakermagazine.com, June 15, 2011.

Hamilton, Clive. *The Earthmasters: The Dawn of the Age of Climate Engineering*. New Haven, CT: Yale University Press, 2014.

Harris, Michael. *The End of Absence*. New York: Current, 2014.

Homer-Dixon, Thomas. *The Upside of Down*. Washington, DC: Island Press, 2006.

Horkheimer, Max. *Eclipse of Reason*. New York: Oxford University Press, 2004. First published in 1947.

Hunt, Terry, and Carl Lipo. "Ecological Catastrophe, Collapse, and the Myth of 'Ecocide' on Rapa Nui (Easter Island)." In *Questioning Collapse*, edited by Patricia McAnany and Norman Yoffee, 21–44. New York: Cambridge University Press, 2010.

James, P. D. *The Children of Men*. New York: Knopf, 1992.

Jeffers, Robinson. *The Collected Poetry of Robinson Jeffers*. Vol. 3. Ed. Tim Hunt. Stanford, CA: Stanford University Press, 1991.

Jensen, Derrick. *Endgame*. Vol. 1, *The Problem of Civilization*. New York: Seven Stories Press, 2006.

———. *Endgame*. Vol. 2, *Resistance*. New York: Seven Stories Press, 2006.

Kermode, Frank. *The Sense of an Ending*. New York: Oxford University Press, 1967.

Kurzweil, Raymond. *The Singularity Is Near*. New York: Viking, 2005.

La Rochefoucauld, François, duc de. *Maxims of le duc de La Rochfoucauld*. Trans. Louis Kronenberger. New York: Random House, 1959.

Lasch, Christopher. *The True and Only Heaven*. New York: W. W. Norton, 1991.

Lovecraft, H. P. *The New Annotated H. P. Lovecraft*. Ed. Leslie Klinger. New York: Liveright Publishing, 2014.

Löwith, Karl. *Meaning in History*. Chicago: University of Chicago Press, 1949.

Maddow, Rachel. Interview with James Inhofe. *Rachel Maddow Show*, MSNBC, March 15, 2012.

Marlow, William. *Trees of the Eastern and Central United States and Canada*. New York: Dover, 1957.

Marten, Gerald. *Human Ecology: Basic Concepts for Sustainable Development*. Sterling, VA: Earthscan Publications, 2001.

Mason, Bill, director. *The Rise and Fall of the Great Lakes*. National Film Board of Canada, 1968.

Matlin, Margaret, and David Stang. *The Pollyanna Principle*. Cambridge, MA: Schenkman Publishing, 1978.

McCarthy, Cormac. Interview. *Oprah Winfrey Show*. June 5, 2007. Oprah.com, June 15, 2017.

———. *The Road*. New York: Knopf, 2006.

McCauley, Douglas, Malin L. Pinsky, Stephen R. Palumbi, James A. Estes, Francis H. Joyce, and Robert R. Warner. "Marine Defaunation: Animal Loss in the Global Ocean." *Science* 347, no. 6219 (2015).

McMurry, Andrew. "The Slow Apocalypse: A Gradualistic Theory of The World's Demise." *Postmodern Culture* 6, no. 3 (1996). See http://pmc.iath.virginia.edu/text-only/issue.596/pop-cult.596.

Meeker, Joseph. *The Comedy of Survival*. New York: Scribner, 1974.

Mendelson, Edward. "Transcendental Rites." Interview. *The Baffler*, 27 (2015).

Merwin, W. S. *The Book of Fables*. Port Townsend, WA: Copper Canyon Press, 2007.

———. *The Shadow of Sirius*. Port Townsend: Copper Canyon Press, 2008.

Nietzsche, Friedrich. *Human, All Too Human*. Trans. R. J. Hollingdale. Cambridge: Cambridge University Press, 1986.

Norgaard, Kari Marie. *Living in Denial: Climate Change, Emotions, and Everyday Life*. Cambridge, MA: MIT Press, 2011.

Orr, David. *Earth in Mind*. Washington, DC: Island Press, 2004.

Parfit, Derek. *Reasons and Persons*. Oxford: Clarendon Press, 1984.

Paris Agreement. United Nations Conference of the Parties Twenty-First Session. Paris, November 30 to December 11, 2015. Accessed October 15, 2017.

Pessoa, Fernando. *The Book of Disquiet*. Croyden: Serpent's Tail, 1991.

Purdy, Jedediah. "Imagining the Anthropocene." *Aeon*. Aeon.co/magazine.

Quinn, Daniel. *Ishmael*. New York: Bantam, 1992.

The Road. Dir. John Hillcoat. Dimension Films, 2009.

Robinson, Kim Stanley. "Is It Too Late?" *State of the World 2013*. Worldwatch Institute. Washington, DC: Island Press, 2013.

Ronell, Avital. *The Telephone Book*. Lincoln: University of Nebraska Press, 1989.

Scheffler, Samuel. *Death and the Afterlife*. New York: Oxford University Press, 2013.

Schiller, Friedrich. *Aesthetical and Philosophical Essays*. New York: Harvard Publishing, 1895.

Schopenhauer, Arthur. *Studies in Pessimism*. London: G. Allen, 1913.

Scott, James C. *Against the Grain*. New Haven, CT: Yale University Press, 2017.

———. *Weapons of the Weak*. New Haven, CT: Yale University Press, 1985.

Sharot, Tali. *The Optimism Bias*. New York: Vintage, 2012.

Shepard, Paul. *Nature and Madness*. Athens: University of Georgia Press, 1982.

Sloterdijk, Peter. *Neither Sun nor Death*. Los Angeles: Semiotext(e), 2011.

Sonnino, Lee A. *A Handbook to Sixteenth-Century Rhetoric*. London: Routledge and Kegan Paul, 1968.

Snyder, Gary. *Turtle Island*. New York: New Directions Books, 1974.

Spengler, Oswald. *Man and Technics*. Trans. Charles Atkinson. New York: Alfred A. Knopf, 1932.

Spinrad, Norman. *Songs from the Stars*. New York: Simon and Schuster, 1980.

Stiegler, Bernard. *The Re-Enchantment of the World: The Value of Spirit against Industrial Populism*. Trans. Trevor Arthur. London: Bloomsbury, 2014.

Theroux, Marcel. *Far North*. New York: Picador, 2009.

Thoreau, Henry David. *The Heart of Thoreau's Journals*. Ed. Odell Shepard. New York: Dover, 1961.

———. *The Natural History Essays*. Salt Lake City, UT: Peregrine Smith, 1980.

———. *Walden*. Ed. Jeffrey Cramer. New Haven, CT: Yale University Press, 2004.

Tiger, Lionel. *Optimism: The Biology of Hope*. New York: Simon and Schuster, 1979.

Tiptree, James, Jr. "The Screwfly Solution." In *Out of the Everywhere, and Other Extraordinary Visions*, 53–75. New York: Del Rey/Ballantine, 1981.

Turkle, Sherry. *Alone Together: Why We Expect More from Technology and Less from Each Other*. New York: Basic Books, 2011.

Twain, Mark. *The Adventures of Tom Sawyer*. New York: Oxford University Press, 1996. First published in 1876.

———. *Autobiography of Mark Twain*, vol. 2. Mark Twain Project, October 15, 2017.

Ulam, Stanislaw. "Tribute to John von Neumann, 1903–1957." *Bulletin of the American Mathematical Society* 63, no. 3 (1957).

Vizenor, Gerald. "Aesthetic of Survivance." In *Survivance: Narratives of Native Presence*. Lincoln: University of Nebraska Press, 2008.

Vonnegut, Kurt. *Deadeye Dick*. New York: Delta, 1982.

Wackernagel, M., and W. Rees. *Our Ecological Footprint: Reducing Human Impact on the Earth*. Gabriola Island, BC: New Society Publishers, 1996.

Wells, H. G. *Mind at the End of Its Tether*. London: William Heinemann, 1945.

Wilde, Oscar. *The Picture of Dorian Gray*. Mineola, NY: Calla Editions, 2016. First published in 1890.

Williams, Joy. *Ill Nature*. New York: Vintage, 2001.

Winters, Ben. *Countdown City*. Philadelphia: Quirk Books, 2013.

———. *The Last Policeman*. Philadelphia: Quirk Books, 2012.

Zapffe, Peter. "The Last Messiah." *Philosophy Now* (March/April 2004): 35–39. First published in 1933.

Index

Adorno, Theodor, 18, 27
albatross, 44
Albrecht, Glenn, 167–68. *See also* solostalgia
Alien, 42
animals. *See specific animals*
Anthropocene, 17, 42
antisingularity, 132–33. *See also* singularity
apocalypse, 1–5, 18, 76, 83, 102
ant, 40
Arctic, 39, 78, 177
Aristotle, 108, 188
arrogance, 31, 48, 71, 144, 160
asteroid, 11, 62–68, 71
atmosphere, 12, 27, 117, 145
Attawandaron, 171, 179

Baker, Carolyn, 108–9
Bashō, Matsuo, 167
bass, 110
bat, 39
Bauman, Zygmunt, 27
bear, 47, 79, 142, 175–76, 179
Becker, Ernest, 69
Beckett, Samuel, 90
Benjamin, Walter, 59
biosphere, 23, 90, 103, 105, 117, 118
birds, 39, 44, 77, 137, 142, 148, 151, 190. *See also specific birds*
bison, 22
blue heron, 188, 190
blue jay, 166–67
Book of Revelation, 76, 107
Bush, George W., 91, 123
Bushmen, 25, 61

Calhoun, Haystacks, 96
Calinescu, Matei, 81–2
Campbell, John W., 44
cape buffalo, 156
capitalism, 12, 59, 87
Captain Marvel, 8
carbon, 12, 30, 39, 89, 95, 105, 132, 149, 160 carbon dioxide (CO_2), 27, 71, 131
Cassandra, 5–6
cat, 127, 176
catastrophe, 4, 5, 43, 45, 67, 72, 80, 90, 103, 110
cheetah 39
chicken, 59, 119, 120
Children of Men, 39, 63, 84–87, 91
Cicero, 68
Cioran, E. M., 165
civility, 31, 32, 103, 122
civilization, 17, 24, 25, 31, 41, 46, 62, 64, 65, 82, 107, 109, 128, 187, 189; collapse of, 92; definition of, 60; destruction of, 66, 96, 153; false god of, 32; global, 89; industrial, 16, 59, 77, 90, 92, 103, 106, 168; march of, 4; removal of, 96–100, 104–5; suicide by, 110, 169; survival of, 49, 56; Western, 37; zenith of, 10, 101
climate, 11, 22, 89
climate change, 5, 6, 48, 57, 64, 89, 122, 131, 156, 167
Clinton, Bill, 15, 78, 91
coal. *See* hydrocarbons
cobra, 176
cod, 12
Cohen, Tom, 42
collapsarian, 77, 108
collapse, 23, 38, 39, 41, 70, 80, 82, 83, 85, 90, 96, 118; biospheric, 43; conscious, 103; environmental, 66, 132; of industrial civilization, 92; principle, 17; slow-motion, 108
computer, 29, 44, 48, 99, 117, 125, 131, 132, 169, 170
coral reef, 30, 115, 157
Costa, Rebecca, 4
crane, 139
crisis, 43, 46, 47, 53, 57, 79, 110, 111, 117; existential, 67; fertility, 63
crocodile, 120, 156

Cthulhu, 15, 76
cynicism, 16, 53, 127

decivilization, 96
decivilizationists, 97, 108, 112
denial, 6, 31, 39, 59, 68, 69, 70, 72, 110, 111
dog, 125, 139, 142, 176, 182, 183
doom, 24, 26, 40, 72, 79, 109, 156; and gloom, 31, 39, 93; impending, 67; saying, 89; time, 2
doomsday, 80, 81, 83, 87, 105, 107
Dos Passos, John, 165

eagle, 163, 179
ecology, 29, 30, 49
ecosystem, 24
ecosphere. *See* biosphere
Edelman, Lee, 84
Eder, M. D., 95
education, 32, 43, 49, 53, 55, 119
Ehrenfeld, David, 48, 54
elephant, 30, 140, 156, 182
Emerson, Ralph Waldo, 40, 188
enlightenment, 4, 25, 27, 29
environment, 4, 12, 48, 60, 63, 98, 130, 160
environmental catastrophe, 4, 43
environmental dread, 39
environmental humanities, 45,
environmentalists, 5, 45, 185
environmental psychopathology, 151
environmental studies, 54, 151
extinction, 2, 28, 39, 41, 43, 63, 66, 73, 80, 110, 131, 132, 156, 170, 179

Facebook, 115, 158
Fernández-Armesto, Felipe, 60, 61, 62, 97
Fitzgerald, F. Scott, 90
forest, 22, 39, 90, 98, 119, 151, 153, 162–63, 174, 179, 182
Franzen, Jonathan, 138
Freud, Sigmund, 82
frog, 40, 99, 125, 182, 185

gasoline. *See* hydrocarbons
gemsbok, 156
glacier, 115, 168, 179
global warming. *See* climate change

God, 8, 9, 15, 75, 77, 124, 143, 144, 149, 151, 152
Gore, Al, 5, 6, 43, 89
gorilla, 24, 29, 83
Gramsci, Antonio, 16
greed, 9, 89, 109, 160
growth machine, 77, 93, 186
Gulf Stream, The, 79

Hamilton, Clive, 11
Hansen, James, 5
happiness, 14, 63, 116
Heidegger, Martin, 133, 189
Homer, Winslow, 79
Homer-Dixon, Thomas, 90
Homo colossus, 26
Homo sapiens, 3, 38, 168
hope, 11, 15, 33, 34, 41, 53, 67, 68, 77–93, 111, 160, 165; delusion of, 67, 77, 158
Horkheimer, Max, 18, 27, 29, 30, 32
horse, 119, 153
human syndrome, 71, 73
humanist, 37–47, 53, 54
humanities, 37, 41–48, 52–54
humans, 11, 25, 26, 38, 83, 86, 109, 127, 132, 139, 149, 151, 155, 176; arrogance of, 30, 37, 137, 141; ignorance of, 145, 185; greed of, 98; impermanence of, 2–4; self-absorption of, 9, 121, 129, 138, 144; stupidity of, 18, 22
hydrocarbons: coal, 12, 71 89, 96, 101, 140, 158; gasoline, 71, 95, 149; oil, 6,12, 71, 80, 101, 105, 118, 139, 140, 152, 153, 169, 179

indigenous peoples, 25, 99, 112 171, 177, 179, 185
ignorance, 49, 56, 71, 111, 156, 189
information, 30, 44, 56
Inhofe, James, 48
insect, 160–61
IPCC, 5

Jackson, Jesse, 78
James, P. D., 63
Jeffers, Robinson, 69, 96
Jensen, Derrick, 68, 98–108
Jesus, 40

Kahn, Herman, 116
Kermode, Frank, 76
Klee, Paul, 60
Kurzweil, Raymond, 131–33

lake, 31, 147–48, 163, 168, 179; Erie, 184, Huron, 183; Michigan, 143, 185
Lasch, Christopher, 16
Last Policeman, The, 65–66
Leavers, 24–26
Leibniz, Gottfried Wilhelm, 14–15
lion, 153, 156
Lovecraft, H. P., 17
Löwith, Karl, 82
Lucifer, 7–8
Lyell, Charles, 1

Maddow, Rachel, 48
Marx, Karl, 46, 77, 82
Malthusian, 23
market, 25, 31, 48, 49, 51, 52, 53, 55, 82, 110, 123
Mason, Bill, 168
McCarthy, Cormac, 84–85, 88
McFerrin, Bobby, 110
McKibben, Bill, 89
McPherson, 108–9
media, 6, 52, 55, 56, 116, 117, 120, 124, 130, 134, 173, 174, 183
Melancholia, 66–68
Mendelson, Edward, 116
Mennonite, 26, 171, 180, 187
Merwin, W. S., 117–18, 184
Michigan, 145, 172, 181
Michigan, Lake, 143, 185
Milton, John, 7
minnow, 182
Modernity, 13, 25, 26, 32, 60, 82, 166
money, 16, 25, 27, 34, 56, 78, 145, 168
monkey, 29, 31, 83
monkey wrench, 100

narcissism, 9, 69, 144
nature, 19, 23, 29, 119, 120, 156, 186, 188; domination of, 12, 27, 30, 32, 60, 71, 116, 166, 167, 185; forces of, 22, 40, 134, 145, 168; harmony with, 25, 31, 98, 100, 109, 137, 174

Nietzsche, Friedrich, 77
Norgaard, Kari Marie, 6
nuclear, reactor, 101; blast, 2, 40, 138

Obama, Barack, 15, 78, 89, 107
ocean, 26, 30, 39, 41, 59, 79, 84, 92, 100, 105, 110, 149, 151, 152, 153, 160, 185, 189
Oedipus, 6–7, 82
oil. *See* hydrocarbons
Ontario, 169, 178, 184, 185, 187
Oprah (Winfrey), 15, 78, 88
optimism, 13–17, 27, 37, 59, 69, 70, 77, 87, 88, 104
orangutan, 139
Orr, David, 52

Paleolithic, 32
Pangaea, 97
Pangloss, 17, 88,
parakeet, 170
Paradise Lost, 7
Paris Treaty, 161–62
Parfit, Derek, 61–64
pessimism, 15–17, 37, 38, 77, 81
Pessoa, Fernando, 22
pig, 142, 175
pigeon, 179
Planet of the Apes, 83–84
Plato, 40, 177
Pleistocene, 133
polecat, 119
Pollyanna, 13–16, 72, 93
progress, 2, 16, 24, 31, 38, 52, 70, 82, 89, 90, 95, 109, 121, 127, 131–33

Quintilian, 92
Quinn, Daniel, 21, 24

rabbit, 139, 175
rat, 138
resilience, 17, 112
rhetoric, 4, 7, 18, 50, 55, 88, 90, 91, 92, 177
rhinoceros, 39
river, 59, 99, 116, 150, 152, 153, 154, 184, 185, 186; Conestoga, 187, 188, 190; Grand, 184, 185, 187; Mississippi, 158, 185; Nile, 31
Road, The, 39, 84–87, 91
Robinson, Kim Stanley, 132, 157

Sagan, Carl, 1
salmon, 22, 92, 94, 99, 101
Satan. *See* Lucifer
Sawyer, Tom, 12
Scheffler, Samuel, 63–66
Schiller, Friedrich, 65
Schopenhauer, Arthur, 16, 37
Scott, James C., 111
shark, 79, 96
Sharot, Tali, 88
Shazam!, 8, 10
sea. *See* ocean
sea slug, 147
Sentinelese, 25
Shepard, Paul, 71
Simpson, Homer J., 15
singularity, 131–32. *See also* antisingularity
Sloterdijk, Peter, 13
smartphone, 121–22, 128, 130, 172, 174, 175, 179, 186
snake, 176
snakehead, 148
springbok, 156
Snyder, Gary, 28
solastalgia, 168–69
Sonnino, Lee A., 92
Spengler, Oswald, 16, 38
spider, 160–61
Spinifex, 25
Spinrad, Norman, 72
starling, 44
Stiegler, Bernard, 133
stupidity, 109, 168
Sunny Sparkle, 8–9
Superman, 8
Supremes, 5–18
survivance, 112
Susenbrotus, Johannes, 92
sustainability, 18, 49, 54, 55, 60, 98, 107, 157
sustainable development, 26
Suzuki, David, 185, 186
Swift Jr., Tom, 10–11
system, 13, 17, 42, 50, 66, 76, 88, 144; complex, 17, 96, 105; natural, 24, 60, 89, 99; semiotic, 38, 44; social, 30, 47, 104, 106; world, 67, 129

Takers, 24–26
technofix, 27, 57, 83

texting 125–28, 173, 174
Theroux, Marcel, 64
Thoreau, Henry David, 31, 32, 149, 189
Tiger, Lionel, 27, 69
Tiptree Jr., James, 84
Toffler, Alvin, 116
topophilia, 189
transcendence, 71, 81, 133, 138
tree, 24, 34, 139, 145, 148, 150, 152, 153, 171, 175, 178, 179, 181; beech, 155; boreal, 132; butternut, 194, 195; cedar, 153, 181; chestnut, 179; elm, 179; hemlock, 162–63; logic, 125, 127; maple, 145, 172, 184; oak, 61, 99, 153, 181; palm, 107; peanut, 140; pine, 48, 115, 162–63; poplar, 61; redwood, 99, 163; spruce, 150, 181; willow, 181, 182, 184
Trilling Lionel, 188
trout, 148, 182
Trump, Donald J., 15, 107, 169; Don Jr., 156–57; Eric, 156–57; Ivanka, 159
Trumpocene, 169
Tuan, Ye Fu, 189. *See also* topophilia
Turkle, Sherry, 124–25
turtle, 182
Twain, Mark, 31, 144
Twitter, 116, 12, 125, 127, 129
Tyson, Neil deGrasse, 1

Ulam, Stanislaw, 131
university, 34, 37, 43, 49–52, 55–57, 151, 176, 178, 186

virus, 40, 52, 96
Vizenor, Gerald, 112
Vonnegut, Kurt, 145
Von Neumann, John, 131
Von Trier, Lars, 66, 67
vulture, 96

water, 11, 73, 79, 83, 96, 98, 100, 104, 105, 118, 139, 140, 143, 148, 150, 152, 153, 168, 182, 185, 188, 190
Waterloo (Ontario), 169, 172, 180, 181, 184, 187; University of, 51
Wells, H. G., 80, 155–56
whale, 30, 98, 163

Wilde, Oscar, 60
wildebeest, 156
wilderness, 31, 32, 134, 152, 185
Williams, Joy, 23
Williams, Roger, 178
Winters, Ben, 65–66
wolf, 3, 28, 96, 179
wolfman, 33
Worldwatch Institute, 157
worm, 30, 48, 93, 99, 119, 154

Zapffe, Peter, 75
zebra, 156